EQUAL NATURES

SUNY series, Studies in the Long Nineteenth Century
—————
Pamela K. Gilbert, editor

EQUAL NATURES
Popular Brain Science
and Victorian Women's Writing

Shalyn Claggett

Published by State University of New York Press, Albany

© 2023 State University of New York

All rights reserved

Printed in the United States of America

No part of this book may be used or reproduced in any manner without written permission. No part of this book may be stored in a retrieval system or transmitted in any form or by any means including electronic, electrostatic, magnetic tape, mechanical, photocopying, recording, or otherwise without the prior permission in writing of the publisher.

For information, contact State University of New York Press, Albany, NY
www.sunypress.edu

Library of Congress Cataloging-in-Publication Data

Name: Claggett, Shalyn R., author.
Title: Equal natures : popular brain science and Victorian women's writing / Shalyn Claggett.
Description: Albany : State University of New York Press, [2023] | Series: SUNY series, studies in the long nineteenth century | Includes bibliographical references and index.
Identifiers: LCCN 2022037736 | ISBN 9781438493152 (hardcover : alk. paper) | ISBN 9781438493176 (ebook) | ISBN 9781438493169 (pbk. : alk. paper)
Subjects: LCSH: English literature—Women authors—History and criticism. | English literature—19th century—History and criticism. | Phrenology in literature. | Literature and society—England—History—19th century.
Classification: LCC PR115 .C48 2023 | DDC 820.9/928709034—dc23/eng/20221017
LC record available at https://lccn.loc.gov/2022037736

10 9 8 7 6 5 4 3 2 1

For my mother and father

Horse was the phlegm of the snicker—in separating the female from her womb.

—Norman Mailer, *The Prisoner of Sex*

Nature does not define woman: it is she who defines herself by reclaiming nature for herself.

—Simone de Beauvoir, *The Second Sex*

Contents

List of Illustrations — xi

Acknowledgments — xiii

Introduction — 1

Chapter 1   Feminist Phrenologists and the Battle for the Brain — 23

Chapter 2   Of Two Minds: Charlotte and Anne Brontë's Use of Innate Psychology — 59

Chapter 3   Harriet Martineau's Material Rebirth — 99

Chapter 4   Mary Elizabeth Braddon's Physiological Critique of Social Identity — 129

Chapter 5   George Eliot and Biological Destiny — 165

Afterword   Battle for the Brain Redux: Brain Imaging, Neurosexism, and Feminist Science — 203

Notes — 211

Works Cited — 231

Index — 249

Illustrations

Figure I.1	"Phrenological Chart—Taken by Mrs Hamilton."	3
Figure 1.1	Phrenological head.	32
Figure 1.2	Table of phrenological organs.	46
Figure 1.3	"Mrs. James Fairman."	55
Figure 2.1	Illustration of the head of Pope Alexander VI.	84
Figure 2.2	Record for a child's daily duties.	89
Figure 5.1	Advertisement for phrenological examinations at Lorenzo Niles Fowler's Phrenological Rooms at Ludgate Circus, London.	171
Figure 5.2	Phrenological chart with portraits of historical figures and illustrations of skulls exhibiting racial characteristics.	179

Acknowledgments

My warmest thanks go to Jay Clayton for guiding this project in its earliest stages and remaining a source of continual support and invaluable advice. I am also grateful to Carolyn Dever, Mark Schoenfeld, and Sander Gilman for helpful discussions and critical commentary that significantly improved the book's shape and direction. I also wish to acknowledge the support of my early teacher Robert Mielke, whose intellectual guidance has had an immeasurable impact on my life and work, and to whom I will remain forever grateful.

This project was made possible by several grants and fellowships from Vanderbilt University, as well as the staff and resources at the National Library of Scotland, the British Library, the Wellcome Library, and the Cadbury Research Library at the University of Birmingham. An earlier version of chapter 3 appeared under the same title in *Victorian Literature and Culture*, vol. 38, no. 1, 2010, pp. 53–73, and a shortened version of chapter 5 appeared in *SEL Studies in English Literature 1500–1900*. Both are reproduced with permission. I also presented early material from *Equal Natures* to engaged audiences at conferences hosted by the Modern Language Association, the North American Victorian Studies Association, INCS—Interdisciplinary Nineteenth-Century Studies, and the International Society for the Study of Narrative. I am thankful for the suggestions and illuminating insights I received from these scholarly communities over many years. I am also indebted to the series editor, Pamela Gilbert, for her initial interest and enthusiastic support of the book. I am particularly grateful to my editor at SUNY Press, Rebecca Colesworthy, who expertly steered this book to publication, and Aimee Harrison for her work on the cover design. Thanks are also due to Lauren Adams, my fantastic research assistant.

This is a book about extraordinary women, and I am profoundly aware of the inspiration and reassurance I have received from the many incredible women in my own life. *Equal Natures* would not have been possible without the emotional support of Hannah Brune, Leah Sheidt, Lauren Sweet-Schuler, Elena Gibson, and Caroline Ford. I am also grateful to my colleagues and friends at Mississippi State University, particularly Devon Brenner, Alicia Hall, Lisa Brandon McReynolds, Holly Johnson, Lara Dodds, and Amanda Clay Powers (now at the Mississippi University for Women). I also wish to thank the members of my writing group—Diana Bellonby, Nora Gilbert, Deanna Kriesel, and Jill Rappaport—whose commiseration and encouragement got me through the final stages of writing as well as a pandemic.

Finally, I wish to express my appreciation for my family. My husband, Matt Lavine, has supported me at every level with infinite patience and sympathy, and I could not have finished this book without him. Lucy Canessa, who is as dear to me as a sister, has been the most enthusiastic champion of this book since its inception, and I count having her in my corner as one of the greatest gifts of my life. Most of all, I want to thank my parents, Sam and Sherrie Claggett, for their boundless love and unconditional support. I dedicate this book to them.

Introduction

In April of 1834, an itinerant lecturer and reformer named Mrs. Hamilton delivered a speech on women's rights to a crowded audience at the Unitarian chapel in Greenock, Scotland. After folding a handkerchief into the "form of a brain," she argued that a correct understanding of that organ would reveal both the source of women's oppression and how they would achieve liberation from their male oppressors. According to one witness, she made the following claims:

> all the bad thoughts, words, and actions of mankind were produced from external impressions, made through the medium of the eyes, the ears, and the other organs of the senses, and that all the errors and ignorance, faults and follies of women, were caused by their being exposed to the foul and contaminating moral influence of bad men; and . . . women's brains were capable of being improved to a degree which would make them equal and even excel the men in all the better accomplishments of our common nature, and give them power to break the chains of the tyrant and the oppressor, and set them completely free. ("Mrs. Hamilton" 32)

For Mrs. Hamilton, the female brain was the ultimate site of conflict and contest: situated at the intersection of external influence and internal capability, it materially encoded the negative effects of male dominance and control and, in doing so, revealed how such treatment was an abuse of nature. Organic structure showed that women possessed brains equal in capability to men, but subjugation literally deformed women's minds, impeding their natural development. Of course, challenging sexual inequality with the argument of social determination was nothing new—Mary Wollstonecraft had made the same point in her *Vindication of the Rights of Woman* more than forty years earlier. Hamilton's appeal, however, superadded the physiological evidence of women's innate mental potential. In this new formulation, nature does not negate or deny the power of external influence to affect women's mental development but rather reveals that the quality of the nurture they receive is incommensurate with their inherent capabilities. Whereas supporters of women's rights had earlier argued that a lack of education prevented the world from knowing what women's natural abilities actually were, this scientific argument went one step further, insisting that women's impressive intellectual capacity could *already* be observed.

Decades later, the astronomer John Herschel and physiological psychologist George Henry Lewes also recognized the implications of brain research for women's rights in a jointly written article for *The Cornhill Magazine*.[1] Responding to recent debates among anatomists about the average difference in size between the male and female brain, they observe, "Let women have the same advantages as men, it is said, and they will exhibit their intellectual equality. Of course there could be no sustaining such an argument if it were demonstrated that women *were* organically inferior to men" (276). Although Lewes and Herschel conclude that neuroanatomy had not advanced far enough to settle the matter, for some Victorian women the brain already offered the proof of possibility for women, and the idea of providing women with all the advantages men possessed was no longer a gamble, but a sure thing.

Unlike Herschel and Lewes, Mrs. Hamilton was not a member of the scientific establishment, and the brain science to which she referred was not neuroanatomy or physiological psychology (both of which would not emerge until the 1860s), but rather phrenology, the popular science of character that held that the shape of the skull was an index of an individual's attributes of mind. Mrs. Hamilton claimed that "phrenologists had proved" that women's minds were equal to men—and she made this argument from authority as she was, herself, a professional phrenologist of considerable renown (32). According to the prominent American phrenologist Lorenzo Niles Fowler, she was the "most successful and correct itinerant lecturer and examiner" in Great Britain, who "travelled extensively . . . and made many friends for herself and for the science" (5). Mrs. Hamilton was one of many professional female phrenologists, in addition to an impressive number of women who wrote manuals and books about the science (Figure I.1).

PHRENOLOGICAL CHART.

TAKEN BY MRS. HAMILTON, PHRENOLOGIST, LONDON.

(LATE OF EDINBURGH) 294, REGENT STREET, LONDON.

	INCHES.		INCHES.
AVERAGE SIZE OF THE MALE HEAD	22	AVERAGE SIZE OF THE FEMALE HEAD	21
Circumference of the head		Anterior Lobe, or Intellectual Region	
From Occipital Bone to Individuality		Coronal, or Moral Region	
From Destructiveness to ditto		Posterior ditto	

THE DOME OF THOUGHT,	MAN, KNOW THYSELF!
THE PALACE OF THE SOUL.	THEN OTHERS LEARN TO KNOW.

"*It is heaven upon earth to have Man's mind to move in charity—rest in Providence—and turn upon the poles of truth.*"

Cerebral Development of J. B. K.

RELATIVE PROPORTION OF THE ORGANS:—Very Large, 20; Rather Large, 18; Moderate, 16; Small, 12; Very Small, 10.
NOTE.—The Organs that are large are those that are naturally most powerful, and which may be most easily cultivated, and in some instances need to be restrained.

DOMESTIC AFFECTIONS.

19 1. *Amativeness.*—Marriage; love. "A man shall leave his father and mother, and cleave to his wife."
18 2. *Philoprogenitiveness.*—Love of offspring and animals.
17 3. *Concentrativeness.*—Love of home; attachment to particular objects and places.
19 4. *Adhesiveness.*—Attachment, friendship, and social sympathy.

PROTECTING FACULTIES.

18 † The Love of Life.
13 † Appetite for Food.
18 5. *Combativeness.*—Courage to meet danger; inspired with boldness to overcome opposition.
19 6. *Destructiveness.*—Gives energy in overcoming difficulties, &c.
18 7. *Secretiveness.*—Prudence in not giving utterance to ideas until judgment has approved of them.
19 8. *Acquisitiveness.*—Industry; a desire to possess articles; provision for old age.
20 9. *Constructiveness.*—To construct houses, machinery, and furniture.

MORAL REGULATING FACULTIES.

18 10. *Self Esteem.*—Inspires the mind with confidence and independence; gives dignity to character; self respect.
20 11. *Love of Approbation.*—Respect for a good name; will inspire to the acquisition of honourable fame.
19 12. *Cautiousness.*—Circumspection before commencing any undertaking.
20 13. *Benevolence.*—Sympathy and charity. "I was hungry, and ye gave me meat."
17 14. *Veneration.*—Respect for the good and great, and revealed truth.
17 15. *Firmness.*—Decision of character.
19 16. *Conscientiousness.*—The sentiment of equity.

IMAGINATIVE POWERS.

17 17. *Hope.*—Keeps up the spirits under misfortunes; inspires the belief in a better state of human existence.
20 18. *Wonder.*—Desire of novelty; admiration of the grand.
20 19. *Ideality.*—Poetic taste; produces the sentiment of beauty.
17 20. *Wit.*—Quick perception of the meaning of others; presence of mind.

OBSERVING FACULTIES.

19 21. *Imitation.*—To imitate virtue; copies the manners of others; necessary to the artist.
11 22. *Individuality.*—Memory for names; produces a desire for the knowledge of objects.
20 23. *Form*—Observes and recollects forms of persons and shapes; necessary to mechanics.
19 24. *Size.*—Gives the idea of space; is essential to the landscape painter and land surveyor.
19 25. *Weight.*—A knowledge of the law of gravitation; essential to genius for mechanics.
17 26. *Colour.*—Perception of colours, and the power of distinguishing shades.

KNOWING FACULTIES.

19 27. *Locality.*—The power of recognizing places; confers a talent for geography; a desire for travelling.
17 28. *Number.*—A talent for arithmetic and calculation.
19 29. *Order.*—Taste for arrangement; neatness in dress.
20 30. *Eventuality.*—A memory for passing events and occurrences; facts of political and natural history.
19 31. *Time.*—The power of judging of the lapse of time; punctuality; essential to the musician.
19 32. *Tune.*—The power to perceive harmony, and relish music.
15 33. *Language.*—Power of acquiring languages, and a readiness in the use of words to express the thoughts of the mind.

REFLECTING FACULTIES.

20 34. *Comparison.*—A tendency to compare one thing with another; traces analogies.
19 35. *Causality.*—Traces the relation between cause and effect, and is not satisfied with analogies, but searches to a Great First Cause.

OBSERVATIONS.

1. Memory depends on the size and activity of any intellectual organ.
2. An organ may be enlarged by exercise.
3. All teachers of youth ought to study Phrenology, as it expands the charitable feelings.

TEMPERAMENT.—*Parts*,—Lymphatic, 20; Sanguine, 6; Nervous, 5; Fibrous,

Figure I.1. "Phrenological Chart—Taken by Mrs Hamilton," mid–nineteenth century, Science Museum, London. *Source*: 4.0 International (CC BY 4.0).

Although phrenology was not created as a science for women, it was nevertheless appropriated by them to advance their personal and collective interests. Most important for this book, however, were the women authors who used this physiological science of mind to question existing social arrangements. Rather than using phrenology as a simple shorthand for characterization or for its iconographic value, these women were deeply concerned with the foundational premise that the basis of mind is physiological rather than solely determined by social influence. In works by Anne Brontë, Harriet Martineau, Mary Elizabeth Braddon, and George Eliot, brain science is a way to think through the implications of an innate identity for such important aspects of personhood as domestic relations, spirituality, public identity, and race. And, much like Mrs. Hamilton, they saw how an appeal to an innatist psychology had the potential to challenge and overturn, at a stroke, ideological assumptions long anchored in tradition and culture.

Equal Natures seeks to restore this lost chapter in the history of women's writing and thinking about science, essentialism, and the body. To be clear, these women writers did not deny the reality of social determination in relation to identity formation; rather, they found its power so prevalent and coercive that they embraced an entirely new foundation for the origin of identity in order to challenge it. Despite women's enthusiastic engagement with the brain's materiality for progressive ends, very little has been said about Victorian women's engagement with brain science. This is most likely because the physiological psychology to which they had access—phrenology—has been debunked and is now invoked as an example of Victorian eccentricity at best, and a racist science at worst. Certainly, phrenology was used throughout the century to support dominant and dominating ideologies in troubling ways, and this book is in no way an attempt to defend a dubious science. The focus here is not phrenology itself but rather on how and why women used an appeal to mental physiology to question entrenched social practices. Although this science was created and popularized by men with little interest in women's rights, the internal logic of the discourse provided women with a point of entry into physiological concepts that, in turn, gave them a biological foundation for challenging established social systems.

This book has three overarching conceptual aims that emerge at the intersection of gender, science, and literary studies. First, it counters the position in feminist theory that essentialism has been the exclusive province of patriarchal values and reactionary political agendas. Essentialism has long been a suspect concept in feminist theory, and rightly so: the history of sexual difference from Plato to present has repeatedly returned to the supposedly irrefutable evidence of nature to justify social and political inequality. In response, the feminist perspective on essentialism has, with few exceptions, remained the same: in 1949, Simone de Beauvoir proclaimed that "one is not born, but rather becomes, a woman" (283), and forty years later Judith Butler similarly observes that "gender operates as an act of cultural *inscription*" (146). Feminist thought has so closely associated itself with social construction that even contemplating the possibility of assuming an essentialist stance seems risky. As Ann Rosalind Jones puts it, "if we argue for an innate, precultural femininity, where does that position . . . leave us in relation to earlier theories about women's 'nature'?" (255). Essentialism, especially in its biological aspect, is the can of ontological worms we dare not open, even if there might be something useful inside.

One exception to this tendency has been in considerations of "the body," which has remained a subject of concern since the 1990s, although it first appeared in 1970s French feminism. Rejecting the hierarchical implications of the Western mind/body dualism that associated women with the body and men with mind, theorists including Luce Irigaray, Hélène Cixous, and Julia Kristeva celebrated women's connection to biological femininity as an aspect of *l'écriture féminine*, which they claimed offered a distinct but equally valuable contribution to culture. This school is unique in the history of feminist thought for adopting a politically and philosophically positive stance toward the body, as most work since—whether theoretical, historical, or literary—has argued that cultural perceptions of sexual difference contribute to the control of women's bodies. Susan Bordo, for instance, traces how internalized sexism materially manifests itself through self-imposed body modifications, and Hortense Spillers has argued that redirecting attention to the material body brings into focus how power marks and disciplines subjects at the intersection of race and gender.[2] Similarly, in her attempt to "rescue the body" from feminist neglect, Elizabeth Grosz has argued that the body is "inscribed, marked, engraved, by social pressures," which necessarily contests the precultural status ascribed to it by the natural sciences (x). These more recent treatments of the body, however, are actually more tempered versions of the constructivist position: they eschew talking about the body as a postmodern abstraction but do not abandon the idea that sexual identity materializes as such through cultural perceptions. As Spillers puts it, to claim otherwise "would appear reactionary, if not dumb" (66).

The project of *Equal Natures* is not to oppose or criticize these important interventions but rather to show that appeals to essentialism and the body have historically been used by women in ways as yet unaccounted for by contemporary critical and theoretical discourse. In part, this is because what all of the approaches outlined above have in common is a focus on sexual difference: on reproductive function, sexuality, or physical dominance.[3] But what happens if, instead of denying the mind/body schism (as in *l'écriture féminine*) or claiming that gender is a construct that gets attached to the body through discursive practice (Butler et al.), we were to critically embrace a mind/body distinction and assert that while the bodies of men and women are different, their minds are equal, *and the material structure and substance of the brain proves it*. With the exception of the italicized portion, this last sentence is hardly a novel perspective—it is just one that has been notoriously difficult to prove. The history of disproportional achievement has been the primary evidence of intellectual difference, and understandably the earliest political critiques of sexual inequality aimed at showing how the argument by achievement failed to account for environmental, social, and cultural influence. Constructivism is a valuable way to invalidate the premise of this causal argument; an appeal to mental physiology, however, is an argument by another type of evidence altogether. This book reconstructs and examines how women authors crafted progressive arguments from this entirely different premise, one based on biologically determined mental capabilities that revealed an innate essence frequently at odds with its social conditions.

The second aim of this book is to restore to the literary and cultural record a different aspect of women's cultural engagement with nineteenth-century science. Much work on women and physiology in Victorian studies, and science studies more generally, has focused on how women have historically been an object of study *for* science. Historians have shown how scientific discourses have defined women as biologically and psychologically different from (and inferior to) men, and that such assessments stemmed from and reinforced the dominant ideology. For instance, Thomas Laqueur has shown that in the Western tradition, the science of sex from the ancients through the Enlightenment supported a one-sex model in which women were underdeveloped men. This was gradually replaced in the 1700s by a two-sex model that naturalized the separate and distinct social roles of men and women.[4] Ornella Moscucci has demonstrated how the development of nineteenth-century gynecology contributed to characterizing women as domestic and maternal, and thus naturally suited for the private sphere (1–5). And, as Londa Schiebinger has argued, the specificity of biological distinctions did more than just reinforce the separate spheres ideology: it effectively precluded women from pursuing science in ways that had been possible in the past, since their domestic natures disqualified them from participating in the increasingly professionalized realm of science.[5] This exclusion was further reinforced by the fact that formal training in physiology was taught in medical schools that barred women from attending because of the perceived "indelicacy" of the subject matter (Farnes 273). In the Victorian period, when it comes to human biological science and women, the story is one of uniform dominance and exclusion.[6]

However, when we consider the popular fields that have since been dismissed as "pseudoscientific," it becomes clear that Victorian women certainly *were* engaging with theories about the biological body. Since the emergence of the history of science as a discipline, nineteenth-century fields of inquiry formerly regarded as pseudoscientific have been reclaimed as serious subjects for scholarship. As Alex Warwick has argued, pseudoscience is typically defined as either a science representing itself as "true" when it is deliberately false, or a science cast off from the establishment for sociocultural reasons. This bifurcated view, however, ignores the relationship between culture and "what it is possible to represent as science at any particular time" (4). Groundbreaking work by Steven Shapin, Roy Wallis, and Alison Winter (among many others) has shown that Victorian sciences anachronistically assumed to be marginal were in fact central to sociocultural developments in the period.[7] Such work has also shown how emerging scientific disciplines claimed legitimacy by distinguishing themselves over and against less specialized popular sciences.[8] This consolidation of authority took the form of developing gatekeeping procedures (including degree programs, professional societies, and specialized journals), and a concomitant effect of these developments was to formally and forcefully shut the door against women.[9] Women, however, could far more easily participate in popular sciences. One form this took was the development of a parallel track of popular science writing, in which women authors could disseminate scientific ideas for the education of women and children. Such contexts, however, often replicated a division of scientific knowledge along the lines of a separate spheres ideology, with women writers acting as domestic counterparts to professional male scientists.[10] In the case of popular or contested sciences, however, women could sometimes achieve roles on par with male practitioners and even contribute original ideas about their object of study. Thus, women became professional phrenologists, mesmerists, and spiritualists.[11] Admittedly, there were comparatively fewer women than men to do so, but given the context of women's virtual absence in the more elite fields of professional Victorian science, such participation deserves attention—not merely because women were participating but because of what they had to say when they were in those rooms and publishing in those journals.

The elision in scholarship about this participation is particularly problematic when gender becomes a consideration, not only because the popular sciences were the ones to which women had more access, but also because women's engagement with contested sciences offers an important counterpoint to the sexist strain of the evolutionary discourse that dominated the era. Charles Darwin's theory of natural selection was revolutionary, but it was also used to legitimate the patriarchal organization of nineteenth-century society. In *The Descent of Man*, for instance, Darwin claims that the "chief distinction in the intellectual powers of the two sexes is shewn by man attaining to a higher eminence, in whatever he takes up, than women can attain—whether requiring deep thought, reason, or imagination, or merely the use of the senses and hands" (327). Basing his argument on physical attributes rather than disproportionate achievement, Herbert Spencer comes to the same conclusion, asserting that "as certainly as [women] have physical differences which are related to the respective parts they play in the maintenance of the race, so certainly have they psychical differences," and to claim otherwise "is to suppose that here alone in all of Nature there is no adjustment of special powers to special functions" (31). Such statements are characteristic of the illogical thinking and implicit bias that pervaded biological assessments of women in the period. As Cynthia Eagle Russett has demonstrated, showing how women were intellectually inferior and physiologically suited only for domestic roles was one of the greatest preoccupations of Victorian science in general, and evolutionary theory in particular.[12] Placed in the larger framework of natural selection, women's social functions became immutable because they were understood as natural capacities resulting from thousands of years of incremental adaptation.

While phrenology lacked the scientific legitimacy later accorded to evolutionary biology, its comparative inclusiveness enabled ways of thinking about natural law that challenged this mounting body of scientific "evidence" that claimed women were innately inferior in mind. Also, while sexist evolutionary thought posited that it would take millions of years of felicitous selection for women to achieve intellectual parity, phrenological discourse could be used to justify an immediate change in women's conditions. The Scottish educationalist James Simpson, for instance, successfully used phrenology to justify his advocacy for boys and girls receiving the same education, explaining in his *Philosophy of Education*:

> Why should the faculties of females, which are the same as the faculties of males, be deprived of the intellectual food which is intended for them? If the cultivation of these faculties shall elevate the male character, will it not likewise elevate the female, and, through the elevation of the female character, unspeakably benefit society? All the *moral* training proposed for the one sex will be granted to be proper and necessary for the other, but not less is the intellectual. (132)

As Simpson points out, the common educational argument that there should be a sexual division of curriculum, in which women receive a moral education but not an intellectual one, is necessarily challenged by the material existence of intellectual faculties in both sexes. As soon as women's brains were understood to have the same morphology as those of men, reformers were able to challenge their unequal intellectual treatment on biological grounds.

Like Simpson, women phrenologists also placed a great deal of hope in the brain as an organ that had no sex. Rather than science being the discourse that naturalizes sexual difference and justifies unequal treatment, in this case a biological science catalyzed and legitimated ideological critique. In a sense, these Victorian women anticipated the work of Donna Haraway, who has called for women to embrace the "sciences of liberation" made possible by twentieth- and twenty-first-century scientific advancements (8). Victorian women's identification of brain science as potentially progressive, however, moves the assumed timeline of sexual liberation through science back by more than a hundred years and reveals a concealed history of women's strategic use of biological essentialism, a scientific discourse otherwise consistently associated with patriarchal dominance. As this book shows, men were not alone in apprehending the power of scientific naturalism to affect cultural perceptions of gender identity. Victorian women's serious consideration of the physiological basis of mind posits innate psychology as a forceful tool to question and overturn problematic social relations.

Finally, this book seeks to significantly expand our knowledge of Victorian literature that grapples with the social implications of cerebral localization both before and outside of the research of mainstream, professional scientists. Recent studies on the connection between literature and Victorian brain science have tended to focus on the impact of physicians, biologists, and psychologists who either possessed or eventually achieved widespread recognition from the scientific establishment for their groundbreaking theories and discoveries. Rick Rylance's *Victorian Psychology and British Culture, 1850–1880* (2000), for instance, limns the development and cultural effects of the physiological theories of Alexander Bain, Herbert Spencer, and George Henry Lewes—three of the most influential representatives of high-Victorian psychology. Similarly, Nicholas Dames's *The Physiology of the Novel: Reading, Neural Science, and the Form of Victorian Fiction* examines how the work of Lewes and Bain, in addition to E.S. Dallas, helped to establish a "physiological novel theory" in Victorian literary criticism, which attempted to elucidate the ways in which narrative form and reading practices might achieve physiological effects.[13] Focusing on neurology rather than psychology, Anne Stiles's *Popular Fiction and Brain Science in the Late Nineteenth Century* (2012) traces how researches into cerebral localization by such neurologists as David Ferrier, John Hughlings Jackson, and Paul Broca influenced the development of the Gothic romance in the fin de siècle. These important studies into the literary and cultural effects of prominent research on the brain by eminent scientists are certainly worthy of our attention, but it is also important to remember that there is a temporal and gendered specificity to readings that focus on Victorian physiological psychology and neurology. As Dames observes, physiological novel theory was largely owing to "the work of a small set of mid-Victorian *male* figures" working between 1850 and 1880 (9, my emphasis). Similarly, Rylance focuses on the work of a small number of male psychologists, whereas Stiles focuses on cerebral localization experiments performed by male neurologists in the 1860s and '70s. This focus on men is not an elision or a misrepresentation on their part—the key players in developing Victorian physiological psychology and neurology *were* men, and the fact that they were men is simply an effect of the structural sexism of the scientific establishment in the second half of the nineteenth century. Long before the 1850s, however, the idea that the brain was the organ of mind—and further, that its physiological properties in large part determined individual personality—had already come to the attention of the British public through phrenology, the first psychology to claim an innatist foundation.

Literary texts are crucial to understanding the popular acceptance of a physiological basis for identity because they form an important part of the public discourse on science in the Victorian period. As Ilana Kurshan has argued, phrenology was particularly well suited to popularization through literature because both center the act of reading, although in phrenology the text is anatomical rather than literary (35). Phrenologists, in fact, even courted this analogy, frequently using the metaphor of reading in their journals (34).[14] Even beyond this surface correspondence, insofar as human identity is a central concern of fiction and philosophical prose, phrenology offered authors a culturally relevant foray into the social implications of a psychological science. Despite the incredible amount of scholarly work available on literature and mainstream Victorian science, and most particularly on evolutionary theory, the relationship between Victorian literature and popular sciences of the period remains less explored.[15] As Barbara T. Gates observes, this tendency in scholarship has led to a "valorization of eminent scientists and their writing" that "paint[s] a limited picture of Victorian scientific culture, both in terms of what science was and in terms of its audience" ("Ordering Nature" 180). By looking beyond the legacies of the most prominent Victorian scientists, *Equal Natures* restores to view the powerful ways in which a popular psychological science challenged conventional understandings of individual identity and interpersonal relations in ways that were particularly relevant to women writers.

Briefly described, phrenology was a science of character based on the research of Viennese physician Franz Joseph Gall, which encompassed five related claims: first, that the brain is the organ of mind; second, that the brain is composed of an aggregate of mental organs; third, that these organs have distinct, or localized, functions; fourth, that the relative size of any given mental organ corresponds to its power; and fifth, that the shape of the skull serves as an index of the power of the organs underneath. Although today phrenology is generally regarded only in terms of its practical application of reading one's character from the shape of the skull, Gall's doctrine revolutionized conceptions about the mind and its operations. As one historian of science puts it, "Gall was the first to treat mental phenomena as well as the human passions (previously located in the heart and elsewhere) as purely organic problems of neuroanatomy and neurophysiology" (Cooter, *Cultural Meaning* 3). The foundational assumptions about identity that phrenology popularized are now widely accepted: that the brain is the seat of consciousness, that educational interests and professional choices are in part determined by an individual's innate mental abilities, and that discovering what one is predisposed for (or "finding oneself") and bringing that self into alignment with external circumstances is one of life's greatest imperatives.[16] As sociologist Thomas Gieryn observes, "The claim that the brain is the organ of mind was phrenologists' monopoly in the early nineteenth century, though today it is fact for everybody. If phrenologists 'got it wrong' by correlating the size of brain regions with cranial bumps, their other claims pushed science forward by moving the question of mental functioning from metaphysics and epistemology to biology, anatomy, and physiological psychology" (122). This is not to say that phrenology deserves to be valorized in the history of science, but it is to suggest that the cultural diffusion of the idea that there is a biological component to our personalities has an unacknowledged connection to popular understandings of the brain in the nineteenth century.

In Britain, phrenology was nothing short of a sensational phenomenon that captured and held the attention of the public for nearly a century. Phrenology was popularly introduced to England by Gall's former medical student Johann Gaspar Spurzheim, who frequently lectured in the country between 1813 and 1831 (Cooter, *Cultural Meaning* 296). These efforts at dissemination, however, pale in comparison to those of the Edinburgh lawyer George Combe, who, after watching Spurzheim dissect a brain in 1816, made it his life's mission to reorganize society around the principles of the science (108). Combe, a major figure in this book, published more than one hundred works on phrenology and its social applications, including his best-selling *Constitution of Man*, which by 1860 had sold twice as many copies as Darwin's *Origin of Species* would by the end of the century.[17] According to Harriet Martineau, it was the fourth most popular book in the English language, surpassed only by the Bible, *Pilgrim's Progress*, and *Robinson Crusoe* (*Biographical Sketches* 275).

Phrenological devotees amplified Combe's work throughout the British Isles, spreading the science by establishing numerous phrenological journals and societies and publishing thousands of phrenological texts.[18] Proponents of phrenology urged its application in nearly every aspect of public and private life, as evidenced in the titles of such books as *Christian Phrenology*, *Phrenology in the Family*, and *Phrenology: and Its Application to Education, Insanity, and Prison Discipline*.[19] The science's popularity and influence, however, is difficult to estimate from statistical and bibliographic records alone since its omnipresence ushered in an entirely new way of conceptualizing and speaking about personality traits. As the editor of the *British and Foreign Medical Review* remarked in 1840, "the rapid diffusion of phrenological ideas under the cover of ordinary language, and without any reference to their true source, is a proof . . . that the new philosophy is making progress" (Forbes 193). For the contemporary reader, the cover of "ordinary language" often conceals phrenology's pervasive presence in Victorian texts, in which casual references to faculties, organs, and prominent foreheads assume a shared context that can today appear merely descriptive. Far more than being a mere Victorian curiosity, phrenology was everywhere, and it popularized an entirely new way of evaluating one's own identity and the identity of others by foregrounding the mind's organic nature.

The popularization of Gall's claims through phrenology dramatically altered social conceptions about the mind and its operations. Prior to the nineteenth century, psychology was the province of philosophy, so Gall's focus on the brain and his insistence on correlating structure with mental function was the first foray into cognitive science. As George Henry Lewes put it in his *History of Philosophy*, "Every impartial and instructed thinker, whether accepting or rejecting Phrenology, is aware of the immense services rendered to Physiology and Psychology" by Gall, who "rescued the problem of mental functions from Metaphysics, and made it one of Biology" (2: 397–98, 407).[20] Victorian psychologist Alexander Bain claimed that because phrenology claimed that character was "founded in nature," it was "really the first analysis of the mind itself that has anything like a basis to go upon. Phrenology, therefore, is even greater in what it implies than in its more immediate and obvious application to deciphering men's characters by their heads" (24). As Bain recognized, the true importance of phrenology lay in its premise rather than its practice. While it would be an overstatement to assert that phrenology was solely responsible for the cultural shift to recognizing an innatist basis for human psychology, phrenological discourse was nevertheless the first science to make these claims and popularize them.

Phrenology's rapid cultural diffusion was in part owing to its numerous social applications. Phrenologists reasoned that assessing innate mental tendencies could help to better orient the individual to society and provide a reliable basis for adjusting institutional practices to better serve individual needs. As George Combe explains, "until Phrenology was discovered, the nature of man was not scientifically known, and . . . in consequence, very few of his institutions, civil or domestic, were founded on principles accordant with the laws of his constitution" (*Constitution* 28). While private readings might aid the individual in selecting the correct profession, hiring the right employee, or choosing a compatible spouse, phrenological reformers (none more zealous than Combe) envisioned large-scale change, such as educational reform, abolishing the death penalty, and altering the treatment of mental patients. Whether applied to institutional reform or used to guide personal decisions, phrenology offered the promise of reconciling the individual to the external world through an appeal to innatist psychology. Phrenology, it was thought, would lead to individual fulfillment and social progress, achieved entirely through a better understanding of mental functions, capabilities, and limitations.

Phrenology's widespread appeal was also due in large part to its ability to embrace both natural and environmental influence as a way of explaining character development. On the one hand, phrenology took as its starting point the premise that personality was an effect of pre-social organic capacities—that is, an individual enters the world with certain mental predispositions that form the basis of his or her character. On the other, phrenology held that environmental forces played a role in either amplifying or suppressing the behavioral expression of these characteristics. It is perhaps best understood as flexible biological determinism: one's innate capacity could be developed by restraining or cultivating specific mental faculties, but there remained an upper limit to improvement. As a psychological theory, phrenology managed to reconcile the formative qualities of both nature and nurture, providing the individual with a unique biological destiny as well as a sense of personal control over it.

Understandably, scholarship examining how Britons applied phrenology to their own lives has focused on how the science worked to naturalize the dominant political, economic, and social interests of the middle class. Phrenology perfectly supported a theory of individualism that could be reconciled with social stability; it was, in the words of Sally Shuttleworth, "reformist rather than revolutionary," promoting increased economic and social opportunities for the middle class through self-regulation (*Charlotte Brontë* 64). This aspect of the science was similar to the central message of Samuel Smiles's *Self Help*, except for the fact that the starting point for improving one's position in life lay in a physiological assessment of organic talents and deficiencies. The system of applied phrenology was primarily created by, and marketed to, men of the working and middle classes, and it emphatically and repeatedly told these men that they were all unique individuals with natural gifts that destined them for specific professions through which they would find personal fulfillment, success, and happiness.

Phrenology's uses were more nefarious when the subject being assessed was not a normative British individual, but an "other," whether racial, pathological, or criminal. Studies on phrenology's use in classifying specific social and racial types have understandably focused on the science's biological determinism, which came to the fore when the science was used to promote proto-eugenic aims, to justify imperialism, or to identify criminal "types." As one historian puts it, "Phrenology was in essence innatist and typological, believing that human behaviour was the outcome of structures and functions of the mind that were fixed by heredity. From there it was not difficult to see human groups as differently endowed . . . and thereby destined for different roles in the history of human society" (Stepan 23). The existing range of perspectives on phrenology characterize the science and its uses as anything but subversive, and these accounts are certainly not wrong for the demographics they address. However, these broader histories—of scientific professionalism, middle-class labor, and British rule abroad—center on the experiences and interests of men, and simply do not account for the science's uses by women.

Unlike colonial subjects or imprisoned criminals, women were able to access phrenology and deploy its discourse for themselves. As Lucy Hartley has shown, part of the appeal of phrenology was that it did not require a great deal of specialized knowledge and could be practiced by amateurs in the comfort of their own homes, completely "outside the confines of the university or the asylum, the laboratory or the operating theatre" (73). Although women were barred from conducting research in these spaces, they could study and practice phrenology within the domestic sphere. Also, because women could not apply the phrenological knowledge they received about the professions for which they were most fit, they could not benefit from phrenology in the way that men ostensibly could—that is, to improve their own socioeconomic status. For this reason, women's insight into their own innate capabilities and mental capacities revealed the injustice of sexual inequality on biological grounds and thus made phrenology, in their hands, a tool of subversion.

To account for this context, the first chapter of *Equal Natures* establishes how women used this popular brain science to contradict and counteract the intensifying claims about women's intellectual inferiority by evolutionary biologists and anthropologists. In response to the Woman Question, scientists including such luminaries as Charles Darwin, Paul Broca, Thomas Henry Huxley, and Carl Vogt argued that to countenance women's equal access to educational and professional opportunities was a waste of resources because women had evolved to be childlike domestic helpmates. While they based their arguments on the more accepted science of craniometry, which held that the overall size of the skull was an index of intellectual power, women's rights supporters used phrenology to undermine the claim that the mind had a sex. This chapter makes the case that there were three key reasons why female phrenologists appropriated phrenology for progressive purposes. First, unlike most anatomical sciences of the day, phrenology was accessible to women and thus enabled them to study their own bodies for themselves. Whereas male scientists saw in female skulls evidence of the male sex's superiority, female phrenologists found evidence of women's equality. Second, because phrenology posited a theory of innate identity not based on sexual difference, it implicitly (if accidentally) contradicted the separate spheres ideology based on the biological theory of sexual complementarity. Because no structural difference was observed in male and female brains, the organ could not be directly connected to other aspects of sexual identity tied to reproductive function—a lack of distinction women's rights advocates were keen to point out. Finally, and perhaps most importantly, because phrenologists never established a separate methodology for assessing women, and because phrenological paraphernalia were designed with men's unique natures in mind, once women assessed themselves with the same tools, the readings had the collateral effect of revealing that women were biologically qualified for professional roles barred to them. The brain's materiality disclosed that women were not "types" defined by their relation to men but psychologically complex individuals with untapped talents and capabilities that ached to be expressed. Thus, to deny women the opportunity to outwardly exercise their internal identities was not only a failure of justice but also a crime against nature.

While feminist phrenologists found in the science a rationale for challenging their access to the same resources as men, women authors saw that such an understanding of the brain could radically challenge traditionally held views across a range of social domains. If cerebral organization determines what makes a person a unique individual, then all social and cultural interpretations based on the efficacy of environmental influence could be called into question on a scientific basis. What happens if a woman's cerebral organization reveals her to be a genius, but the only roles available to her are as a wife and mother? If character is largely fixed at birth, then how can women's much-vaunted moral influence be either effective or important? How can a materialist view of mind possibly be reconciled with the central tenets of Christian theology and free will? In asking such questions, Victorian women writers used the idea of an innate identity to put pressure on beliefs and practices that seemed to them dangerous, inequitable, or immoral.

The women writers addressed in this book were well versed in phrenological theory and terminology, and most had personal relationships with some of the foremost champions of the science. George Combe was an early mentor of George Eliot and corresponded with Harriet Martineau; Charles Bray, the so-called philosopher phrenologist of Coventry, was close friends with both Mary Elizabeth Braddon and George Eliot; Martineau bequeathed her skull and brain to her close friend Henry George Atkinson, an avid phrenologist and mesmerist. Evidence also suggests that Charlotte Brontë, Martineau, Eliot, and Braddon received phrenological readings on one or more occasions, and Eliot even had a cast made of her skull.

Despite the pervasiveness of this popular science and the degree to which it penetrated the lives of these important authors, there have been no book-length studies on the relationship between phrenology and Victorian literature, with the notable exception of Sally Shuttleworth's *Charlotte Brontë and Victorian Psychology*. Using a Foucauldian framework, Shuttleworth convincingly argues that Charlotte Brontë's application of phrenological theory in her fiction aligns with the dominant ideology of the upwardly mobile middle class. Thus, Charlotte Brontë's use of physiological psychology ultimately reinforces the values of self-discipline and self-regulation that characterize a capitalistic society. In the wake of Shuttleworth's book, Charlotte Brontë has become the most widely recognized literary representative of phrenology in Victorian literature.[21] Charlotte Brontë's use of the science is far from subversive and completely in lockstep with the dominant use of the science by her male contemporaries—but her application is also completely atypical of its use by other notable women writers of the period. In viewing Charlotte Brontë as *the* literary representative of phrenology, we miss seeing the more progressive use of popular brain science by other notable women writers in the Victorian age.

The second chapter makes this pivot by clarifying the distinctions between Anne and Charlotte Brontë's use of phrenology in similar contexts. Whereas Charlotte's references to innate mental capacities in *Jane Eyre* naturalize the meritocratic values of an upwardly mobile middle class, Anne's use of biological determinism undermines and critiques the myth of feminine domestic influence—a belief that perniciously reinforced the separate spheres ideology by rendering legal rights and protections unnecessary for virtuous women. Because a key argument for phrenology's utility was spousal selection, the science's treatment of marriage strongly emphasized the inflexibility of character. By the same logic, however, no recourse existed for those who had entered into marriage while ignorant of their partner's congenital predispositions, leaving women with no legal protection against innate depravity. In *The Tenant of Wildfell Hall*, Anne Brontë dramatizes the infelicitous consequences of such biological incompatibility to morally justify a wife's natural right to leave her husband and retain custody of her child.

Much like Anne Brontë, Harriet Martineau recognized how a physiological approach to human psychology could destabilize the foundational premises of long-established institutions. Moving from marriage to religion, the third chapter examines Martineau's use of popular brain science to justify her belief in materialism and atheism. Her first public rejection of theism appeared in *The Letters on the Laws of Man's Nature and Development*, which she co-wrote with the phrenologist Henry George Atkinson. *The Letters* argue that human consciousness is the product of cerebral localization and therefore has no basis in soul. Although the work has long been dismissed for its frequent references to phrenology and mesmerism, Martineau considered it to be one of her most important and personally significant works. The chapter argues that throughout the *Letters*, Martineau self-consciously appropriates the rhetorical conventions of the confessional narrative form to publicly perform a secular conversion. Her use of the confessional mode mirrors the structure of phrenological conversion narratives, in which the newly awakened devotee recalls a revelatory moment of self-discovery through an illuminating personal reading. In addition to already being popularly linked to a form of secular conversion, phrenology was also at the center of an ongoing debate about the atheism implicit in a materialist approach to mind. By aligning herself with the radical school of materialist phrenologists and adopting a posture of antagonistic provocation, Martineau strove to envision and instigate a radical break with theism through an embrace of biological science. In doing so, she sought to wrest the concept of an essential identity away from a spiritual context and locate it, instead, in the physiological basis of mind.

In the same way the mind's materiality revealed a disjunction between religious doctrine and biological fact, it also made clear the potential misalignment between one's socially ascribed role and innate mental capacities. Chapter four demonstrates that this gap between one's publicly legible identity and biologically fixed psychology was a central concern in Mary Elizabeth Braddon's early fiction. Her first novel, *The Trail of the Serpent*, opens with a head reading of a universally beloved young man named Jabez North whom an itinerant phrenologist reveals to have the skull of a craven murderer. Although the community vehemently rejects the diagnosis, the narrative soon confirms it as North murders a child and states his intent to engage in a series of premeditated crimes. This sensation novel maintains its suspense through the ironic distance it maintains between the physiologically informed but powerless characters who are aware of North's true identity and the institutions that dismiss physiological evidence. Braddon uses the same structure more subversively in *The Doctor's Wife* to cultivate sympathy for a female protagonist whose skull attests to her tremendous intellectual ability that is wasted in tedium when she becomes a middle-class housewife. The narrative trajectory works to close the gap between physiological identity and its external recognition, finally achieved through the *deus ex machina* of her husband's death and an implausible inheritance that frees her to use her innate talents with agency rarely accorded to her sex. In both novels, innate psychology operates as the primary mechanism through which Braddon foregrounds the blind spots of ideological misprision in existing social relations.

Whereas the women writers in the aforementioned chapters valued phrenology for its potential usefulness in promoting progressive ends, George Eliot became increasingly critical of the science. Nevertheless, she remained committed to the foundational principle that both popular and mainstream brain science shared: that the brain is the organ of mind and cerebral organization shapes individual identity. The final chapter charts Eliot's evolving consideration of the role biology plays in determining an individual's future and in envisioning collective destiny. In her early story "The Lifted Veil," she uses an accurate phrenological reading to dramatize the negative psychological effects of being made prematurely aware of one's intellectual capabilities and limitations, ultimately leading to abject passivity and fatalism. Turning from the individual to the species, she satirizes a phrenological reformer in her 1865 poem "A Minor Prophet," whose utopian schemes only thinly conceal proto-eugenic implications. Both works illustrate the ways in which visual signs and tactile measurements of the body tend to delimit future possibilities in troubling ways. Returning to the intersection of the body's materiality and future knowledge in *Daniel Deronda*, Eliot promotes an alternative approach to physiological foreknowledge through Mordecai's visual identification of Deronda in the service of strategically counteracting the coercive effects of racist imagery. Deronda's conspicuous and prepossessing visual presence allows Mordecai to posit him as an emblematic instantiation of essential Jewish identity imbued with the power to unify a people and inspire a future besides either assimilation or subjugation. Although the novel endorses this strategy in a racial context, the narrative treats its application to gender politics more skeptically, revealing the limitations of essentialist appeals for sexual equality within a cultural identity founded on patriarchal practices. Nevertheless, in Deronda's final encounter with his mother, the novel upholds the legitimacy of her radical claim for freedom from sexual bondage on essentialist grounds, rehearsing the same arguments of biological intellectual equality made by the feminist phrenologists discussed in the first chapter of this book.

Undeniably, biological essentialism has been a forceful tool in the hands of the politically and professionally privileged, and more often than not, it has been used to efface the socially constructed origin of inequality by coding it as "natural." Yet, in the Victorian period, women recognized the epistemological force of biological science and appropriated it for their own purposes. As Diana Fuss has observed, essentialism is in itself "neither good nor bad, progressive nor reactionary, beneficial nor dangerous," and thus only the way it is deployed can be ethically assessed (xi). Examining why and how women used brain science to scrutinize and overturn longstanding cultural assumptions about the origin and nature of human psychology and intellectual capabilities recovers an important gambit for power by the disenfranchised. What all the women writers in this book believed is that if social inscription is the tool of oppression and obfuscation, then the best thing to countermand its mark is nature. A poststructural inheritance has long taught us to be wary of foundational claims "centered" on a premise that organizes the system that the premise itself escapes. But long before the deconstructive turn in feminist theory, women used popular understandings of the brain's physiology to criticize and contest sexism, religious dogmatism, and racism. Their strategy did not deny the logic of a center based in nature but, rather, insisted that nature itself testified to the distortions of inequitable systems. *Equal Natures* attempts to demonstrate the merit of this approach which was so ardently and hopefully adopted by women who saw in science a chance for change.

Chapter 1

Feminist Phrenologists and the Battle for the Brain

In the mid-1880s, the influential social reformer and women's rights activist Frances Power Cobbe visited the human skull collection at the Museum of the College of Surgeons. Failing to see a single woman's skull among the vast array of notable male heads, she asked the guide to produce "the best female skulls in the Museum," but he answered that they had only a few unimpressive ones taken from paupers (112). To remedy this unequal representation, Cobbe said she would donate her own skull and subsequently added a codicil to her will bequeathing it to the college. Cobbe related this story in *The Englishwoman's Review* in response to a recently published piece by Charles Darwin's former research assistant, the evolutionary biologist George Romanes, in which he asserted that women were intellectually inferior because the brain weight of the average man outweighs that of the average woman by five ounces ("Concerning Women" 516–18). Suspicious of his evidence and methodology, Cobbe questioned,

> *who* were the women on whose minute and disproportioned skulls Professor Romanes has founded his calculations? [. . .] were they exclusively women of the class who die in workhouses, whose poor minds have been "cabined, cribbed, confined" . . . in the narrowest circle of sordid cares and ideas? Was it *these* women, I ask, whose skulls were measured against those of men, probably of a mixed collection of men,

> including some highly gifted and cultivated ones, and certainly of men having had the advantage of larger interests and better education? ("Feminine Brains" 111)

Provisionally granting the possibility that the size of the brain might be an index of its power, she adds to this the idea that if developed under proper conditions, female brains might supply empirical evidence of innate capability rather than limitation. The skull that she then offered to the museum, as she well knew, was remarkable enough to visually make this point for her. As she explains at the beginning of her autobiography, "It is always amusing to me to read the complacent arguments of despisers of women when they think to prove the inevitable mental inferiority of my sex by specifying the smaller circumference of our heads . . . in my case, as it happens, their argument leans the wrong way, for my head is larger than those of most of my countrymen,—Doctors included." She goes on to list the exact dimensions of her remarkable skull, which she assures her readers was "measured carefully . . . by a skilled phrenologist" (4). Cobbe's response to Romanes was approvingly summarized in *The Phrenological Magazine*, along with the comment that "Perhaps ladies of noted attainments will take Miss Cobbe's hint, and leave their skulls to the Royal College of Surgeons . . ." ("Women's Brains" 233).

In her rebuttal, Cobbe invokes the phrenological dictate that external circumstances and faculty exercise could measurably change the size and shape of the brain. This idea opposed the then predominant view among evolutionists that intelligence was fixed at birth and could not be meaningfully affected by circumstance. Romanes was one of a clutch of anthropologists and evolutionary thinkers—including Carl Vogt, Thomas Henry Huxley, and Charles Darwin—who tried to establish the intellectual inferiority of women on scientific grounds. While this usually took the form of skull measurements, their form of biological determinism lacked the flexibility of Cobbe's, which insofar as it speculatively embraced cranial shape as a possible index of intelligence, also included the contouring effect of social forces on the biological body. Phrenological discourse also did not hold that absolute size was the principle indication of mental capability due to the oft-invoked rule of *cœteris paribus*: "It should be kept constantly in view," George Combe cautions, "that it is the size of each organ in proportion to the others, *in the head of the individual observed*, and not their *absolute size* . . . that determines the predominance in him of particular talents or dispositions" (*Elements* 183). The phrenological system defined the mental capability of an individual in reference to the relative interactions of discrete organs, as opposed to the emerging view in evolutionary anthropology that held only a single trait (that of absolute skull size) accounted for human intelligence. Cobbe also contradicted this assumption, noting that the mathematician Mary Somerville's head was the smallest she had ever seen, demonstrating that a "tiny purse full of gold is better worth than a bag of halfpence" (111). Overall, Cobbe put very little stock in the absolute size of the head indicating intelligence, suspecting instead that the supposed differences between the male and female brain were due more to the selection of the scientific observer than to that of nature, and that the actual evidence would reveal a wide variety of female skulls that would ultimately contradict the limitations and biases of Romanes's putatively objective research.

Romanes and Cobbe represent two opposed interpretations of the brain's materiality in relation to sex: one that claimed intelligence could be reduced to a specific, unalterable trait showing women's intellectual limitations, and another that saw the organ of mind as a product of multiple, mutable properties that revealed women's inherent capabilities. As Stephen Jay Gould has shown, it was the distillation of intelligence to a single, quantifiable criterion that made the practice of hierarchical ranking along a fixed scale possible. This, in turn, led to justifying the denial of privileges and rights to oppressed groups on a scientific basis. Gold omitted phrenology from his study because, unlike other quantitative measurements of mental capability, it did not reduce intelligence to a single measurement (n. 25). This is not to say that phrenology was not used to legitimate oppressive practices—it certainly was, particularly when it came to race. What made phrenology different in the arena of sex had less to do with the science itself and more to do with who had access to it. The evolutionary anthropology that consistently claimed women were inferior based on the overall size of their skulls was also an emerging discipline that emphatically excluded women. But because women had direct access to phrenology and could interpret the evidence of cranial formations for themselves, they drew conclusions that contradicted the findings of the leading biologists and anthropologists. While evolutionary psychology won the battle for legitimacy in nineteenth-century science, the evolutionists' interpretation of women's brains as evidence of innate inferiority was just as erroneous as the assumptions underpinning phrenological assessments of character. At that time, the meaning of the brain's materiality was up for grabs, and whether that meaning was sexist or subversive depended largely on the ideology (and sex) of the interpreter.

In order to understand the maneuverability feminist phrenologists located in their science, it is crucial to make a distinction between phrenology and craniometry, the latter being the chief methodology employed by evolutionary anthropologists in the second half of the nineteenth century to study differences between the races and sexes. Although phrenology readings usually began with a measurement of the skull's general dimensions, the majority of the assessment was a qualitative, descriptive analysis of personality traits accompanied by recommendations on how one could moderate or modify the expression of these aspects. In its practical application, phrenology was a diagnostic psychology aimed at identifying and describing the unique character traits of the individual client. In contrast, craniometry focused on the overall size of the skull in order to provide a numerical analysis of the intelligence of demographic groups. Whether using tape, calipers, or packing shot, this approach sought to establish a biological hierarchy of mental capability supposedly produced through the process of natural selection. In Europe, the method was brought to prominence by French anatomist and anthropologist Paul Broca, best known for discovering the area of the brain linked to speech. Broca collected the most data on the difference in size between male and female skulls, which he interpreted as evidence of men's mental superiority (Gould 103). He also claimed that this difference had increased through time, which he attributed to the species' need for active and rational men; whereas women, as mothers and nurturers, did not need to develop intellectually in order to survive (103–4). In reexamining this data, Gould found that Broca's interpretations were scientifically unsound, and even a product of the misrepresentative selection bias Frances Power Cobbe suspected.[1] At the time, however, this preponderance of quantitative data was embraced by the scientific establishment as incontrovertible evidence of women's intellectual inferiority.

Like Broca, other biologists who turned their attention to women's mental inferiority repeatedly attempted to fold what little knowledge of the brain they had into a larger narrative of evolutionary history. As

Carl Vogt writes in his *Lectures on Man*, "the type of the female skull approaches, in many respects, that of the infant, and in a still greater degree that of the lower races" and "the difference between the sexes, as regards the cranial cavity, increases with the development of the race, so that the male European excels much more the female than the Negro the Negress" (81). Darwin was much impressed by this unsubstantiated argument, which he cites in *The Descent of Man* as evidence of the futility of providing higher education for women as a means of improving the species.[2] Vogt attributed the link between the increase in civilization and the growing intellectual disparity between men and women to the sexual division of labor in more civilized societies (82). In this way, he handily transforms separate spheres ideology into an evolutionary inevitability, and the continued diminution of women's intellects into a sign of biological progress. Broca even warned against any interference into women's position as maternal homemakers on evolutionary grounds: "Everything that affects [the] normal order necessarily induces a perturbance in the evolution of races . . ." (50). Naturalizing women's domesticity and placing it within the context of a progressive evolutionary narrative, the campaign for women's rights became something akin to a pernicious form of artificial selection, impeding the advancement of the species.

For the most part, however, the evolutionary biologists and anthropologists who took up the woman question maintained that the intelligence gap between men and women was so vast that it would be scientifically improbable, if not impossible, for women to catch up with their betters. As Romanes put it, "how long it will take for the woman's movement to evolve the missing ounces of the female brain, as an evolutionist I am afraid to surmise, lest I should be fallen upon with greater fury . . . by the ever-active promoters of that movement [for women's equality]" (Romanes, "Concerning Women" 517). Thomas Henry Huxley similarly observes, "The most Darwinian of theorists will not venture to propound the doctrine, that the physical disabilities under which women have hitherto laboured in the struggle for existence with men are likely to be removed by even the most skillfully conducted process of educational selection" (74). Huxley suggested women not be denied equal opportunity, assuring his audience that they would inevitably fail when pitted against men who were in possession of such "massive brains" (73). His position was relatively liberal compared to most, who tended to agree with Darwin that education and professional training was simply a waste of resources given women's clearly innate mental inferiority.

One of the most vociferous groups to oppose women's rights because of their small brains was the London Anthropological Society, which formed in 1863 in response to the Ethnological Society's decision to admit women (Richards, "Huxley" 255). One of the first motions passed by the society was to officially bar women from meetings under all circumstances (263). Although the central preoccupation of the society was racist ethnography, after the cause of women's suffrage was introduced in 1868 to the House of Commons by the staunch environmentalist John Stewart Mill, they regrouped to render women's equal political treatment a biological absurdity.[3] In an article titled "On the Claims of Women to Political Power," Luke Owen Pike made the connection between the political atmosphere and the science explicit by pronouncing, "I know of no subject upon which [anthropology] ought to give a more authoritative decision than upon the claims of women to political power" (47).

A central argument used by anthropologists to make their case against women's political equality was to assert women's physical and mental likeness to infants, which in turn accounted for their attractiveness to men. In an essay by the German anthropologist and anatomist Alexander Ecker, translated by the Anthropological Society of London and published in their journal, he asserts that the most salient quality of the woman's skull is its resemblance to "the infantile form," which aligns with what is already known of women's characters: that they hold "an intermediate position between man and child" (351, 352). He further observes that the infantile form of the skull in females is particularly prevalent among "especially handsome" heads and concludes that because this shape is most pronounced in heads that men "designate beautiful and womanly, proves that this form is typical for the female sex" (355). In a later article that also cites Ecker and Vogt, J. McGrigor Allan similarly observes that women tend to have an "infantile type of head," adding that women who do not possess small, childlike heads are typically unattractive, as feminine beauty is affected "injuriously by a forehead very high and wide" (204, 205). In other words, when it comes to sexual selection, women are most desired as a mate when they are least intellectual in appearance. This, in turn, suggests that women who might possess an intellectual capacity equal to that of men would not be evolutionarily selected, as their atypical skull renders them naturally unfit. Through this circular reasoning, the male libido becomes a divining rod for the deep evolutionary history of the female of the species: whatever men are most attracted to is the typical form, and the typical form is that to which men are most attracted.

Taken together, Victorian anthropologists and evolutionary thinkers who addressed the question of women's brains were univocal in propounding and supporting the view that the subjugation of women was an evolutionary inevitability. Their reasoning implied that long before humans were cognizant of natural selection or the fact that the brain was the organ of mind, men had been instinctively selecting for certain traits that would ensure the survival of the species, and in women these traits did not include, nor had they ever included, intelligence. By adding an evolutionary paradigm to the biological distinction of women's bodies from those of men, anti-feminist anthropologists and biologists could cast women's subordination as natural, irreversible, and necessary for the survival of the species. Further, this stance rendered women's unequal treatment entirely impersonal, as it was not men's unjust suppression that rendered them inferior (as women's rights activists so frequently claimed) but rather nature's indifferent selection through time. Perhaps most insidiously, however, it suggested that the fight for women's equality could never be won, because it had been lost millennia ago by our primitive ancestors.

These respected men of science, however, were not the only group who saw the strategic value of the brain's materiality for sociopolitical purposes. Women perceived that phrenology offered an entirely different rubric for the interpretation of women's intelligence—one that, precisely because it held the brain was composed of many different organs with relative degrees of power, could not be reduced to the single classificatory trait produced through craniometry. Further, as *the* organ of intelligence, the brain could be conceptually separated from the sexed body, thereby making more apparent the distinction between physical and mental traits. Allan, in fact, pinpoints women's rights activists' attempt to isolate the brain from the biological body and foreground its complexity and power as the catalyst for his vitriolic essay against women's political and social equality: "What is meant by the glib assertion, that woman is the equal of man? Is she equal in size? No. In physical strength? No. In intellect? Yes, replies the advocate; and if she received the same training as man, she would demonstrate her intellectual equality and her moral superiority to her masculine tyrant." The advocate Allan identifies is the women phrenologist and her ilk, who he puns "have 'gone a-head' of equality; and adduce as a proof of the superiority of woman, 'the greater complexity of woman's physical organisation'" (196). As Allan correctly deduced, the organ of mind was the most crucial biological site for establishing the intellectual inferiority, equality, or superiority of women on scientific grounds, not the least because arguments for women's suffrage and education were based on mental, rather than physical, parity. Women phrenologists and phrenological enthusiasts were aware of this too, and they embraced this accessible science to make arguments—both explicit and implicit—for increased freedoms and recognition of women's inherent intellectual capabilities.

While we have far less writing on brain science by Victorian women than we have by men, the available evidence of women's use of phrenology reveals that their engagement with the science was frequently aligned with progressive ideals. This makes sense for three reasons: First, phrenology was one of the few anatomical sciences open to women, allowing them to pursue self-directed studies of their own bodies and to interpret biological evidence for themselves. Second, by offering a theory of innate identity not rooted in sexual difference, phrenology implicitly contradicted the biological foundation of the complementarity of the sexes, which in turn undermined the separate spheres ideology. Third, phrenology provided the basis for a theory of character that allowed women to conceive of themselves as unique individuals qualified for roles outside of the sexually proscribed positions of "wife" or "mother." These aspects of the science were not initially intended for the benefit of women, but this socially constructed science of character that served the interests of those who deployed it was nevertheless appropriated by women to serve their own interests—not, in their case, to control and subjugate others but rather to challenge the legitimacy of social control through an appeal to biological essentialism.

Although they could not initially benefit from phrenology's socioeconomic promises, women in Britain flocked to phrenology lectures and, next to tradesmen, were the largest demographic group to attend (Cooter, *Cultural Meaning* 190). Women's interest in phrenology was so pronounced that it became a source of ridicule by critics of the science. For instance, a correspondent to the *London Medical Gazette* scoffs that rather than garnering the attention of reputable anatomists and physiologists, phrenology appealed to "blue stocking ladies" who "are not idle at home with their mapped casts of heads, and can usually put their fingers upon either wit, merit, adoration, adhesiveness, or philo-progenitiveness, &c. without the least hesitation . . ." (826, 827). An installment of *Blackwood's Magazine*'s popular series "Noctes Ambrosiane" similarly mocks the abundant female phrenologists: when a character observes that "Phrenology is quite epidemic" among women in Edinburgh, the wise "Shepherd" (a character based on James Hogg) replies, "Hae na ye observed that a' leddies that are Phrenologists are very impident, upsettin', bauld amang men, loud talkers . . . grow red in the face gin you happen to contradick them—dinna behave ower reverently to their pawrents, nor yet to their husbands . . ." (Wilson 117, 118). Such characterizations point not only to women's marked interest in the science but also to men's discomfort with this enthusiasm, prompting them to stigmatize such women as unladylike and pretentious.

Possibly contributing to this perception of female phrenologists as willful and independent was the fact that many women distinguished themselves as experts in the science, either by becoming professional phrenologists or by writing books on the subject. Notable practical phrenologists and lecturers include Eliza Sharples (later Carlile), a socialist-feminist radical who came to prominence as a lecturer on social justice in London and also established and edited *Isis*, the only radical journal of the period edited by a woman (Brake and Demoor 570).[4] Ida Mitchell Ellis, a practical phrenologist of Leeds and Morecambe, established the Universal Phrenological Society in 1891, and also edited its associated journal, *Know Thyself* (Cooter, *Phrenology* 122). In addition to these women, many others published books about phrenology, under such titles as *Christian Phrenology*, *Phreno-Physiology*, *How to Improve Body, Brain, and Mind*, and *A Manual of Mental Science for Teachers and Students of Childhood*.[5] Of especial interest, a Mrs. L. Miles published an instructional set of cards on the science, on one of which she observes that "To the female sex, in particular, this science opens a wide field for the exercise of those quick and perceptive faculties with which they are peculiarly gifted . . ." (9). A rarity among phrenological heads (which are almost always unisex), the cover image presents the handsomely embossed profile of Athena—an appropriate choice, as she is both the goddess of wisdom and born from Zeus's forehead (Figure 1.1).

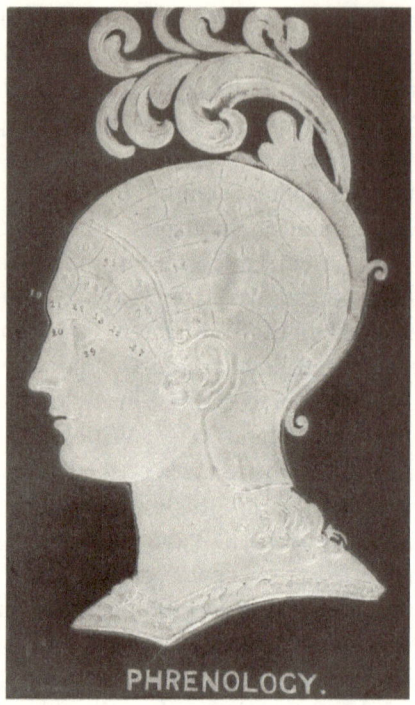

Figure 1.1. Phrenological head from Mrs. L. Miles's *Phrenology and the Moral Influence of Phrenology*, forty pocket-sized cards, [1835]. *Source*: The British Library/Granger.

Aside from being a stylish commercial product, however, using the head of a powerful woman to diagram the placement of mental organs serves as a fitting emblem for what women did with the science: whether motivated by intellectual interest or commercial gain, women saw phrenology as a science in which they could directly participate, so they claimed it for themselves.

As aforementioned, in being available for such appropriation, phrenology was rare among Victorian sciences since at that time increasing professionalization made mainstream science even less accessible to women. As Londa Schiebinger has shown, the separation of public and private spheres that occurred around the same time science became a distinct field helped to ensure that women would not be able to engage in scientific pursuits directly or publicly. Confined to the domestic realm while men were able to seek education and certification at universities, women could pursue science only privately, usually as assistants to men in their family (245–46). Further, and especially in the first half of the century, women were routinely barred from scientific societies, including the Royal Society, the Geological Society, and the Royal Horticultural Society. Although women were initially allowed to participate in the British Association for the Advancement of Science (established in 1831), their attendance was limited to evening gatherings aimed at general entertainment and amusement, with the women themselves serving as an enlivening element for the men (Morrell and Thackray 149). Similarly, although the Ethnological Society briefly allowed women to attend their scientific meetings (a practice Huxley put an end to as soon as he became president), they were seen as serving an ornamental function that would increase the enjoyment of men and therefore lead to increased membership (Richards, "Huxley" 255, 275). Thus, even when women were allowed into the literal halls of science, they were rarely direct participants who were taken seriously by the organizations.

In contrast, women were far more integrated into public phrenological gatherings and were enthusiastic in appreciating their inclusion.[6] The Edinburgh Philosophical Association, one of the earliest institutions to aid in the diffusion of phrenology, even passed a resolution regarding the attendance of women at scientific lectures, declaring, "At these lectures, Females have an opportunity of receiving instruction, which is denied them in nearly every other institution for education [and] have largely availed themselves of the advantages presented to them" ("Summary of Proceedings" 245). Women not only availed themselves of these opportunities but also expressed their awareness that such inclusion was unique. For instance, a group of women presented a pair of silver calipers to George Combe to thank him for being the "first lecturer on a serious subject who had admitted their sex to his class" (Gibbon 208), and, upon Combe's death, *The English Woman's Journal* remembered him as "Woman's Friend" who "recognised no distinction of sexes" and "demonstrated the interest men have in raising the condition and relative station of women" ("George Combe" 53, 55). In her memoir, American physician and women's rights activist Harriot Hunt provides a firsthand account of how phrenology affected women who yearned for acceptance into scientific circles:

> [W]e perused medical works with much dissatisfaction. This probably arose in great measure from our being entirely shut out from the medical world, having no minds with which to interchange views, compare thoughts, and examine experiences. . . . We felt the need of a clear, exploring light: at last we found it. George Combe came to this country, and, in October, 1838, commenced a course of lectures. . . . To me they were revelations—bread for a hungry spirit, and water for a thirsty soul. (142)

In a similar show of gratitude memorializing Johann Spurzheim, a woman writer for the *Ladies Magazine and Literary Gazette* proclaims, "There are reasons which should make my own sex revere his character and be zealous in studying his doctrines. He was the friend of woman. He entertained exalted views of the great benefits which would result to society and the world, from the influence of female intellect, judiciously cultivated and rightly directed" (Saunders, "Dr. Spurzheim" 572). What such accounts reveal is the shared perception among women that the ambassadors of phrenology were unique, both in their recognition and appreciation of women's intellectual capabilities and in their willingness to include women among their ranks.

In truth, male phrenologists (including Combe and Spurzheim) were usually not zealous advocates for women's rights, but the degree to which they wanted to include women eventually became beside the point.[7] Having lost the battle for inclusion and legitimacy within the ranks of the emergent scientific establishment earlier in the century, phrenologists shifted to a strategy of popular appeal to gain power, and numbers were numbers. As Spurzheim himself observes, "From the beginning the fair sex has been favorable to our science. . . . Very few of the medical profession think proper to be interested in our investigations, and prefer dinners and suppers to phrenology. The greater number of the gentlemen are occupied with mercantile speculations; ladies, alone, turn their minds toward scientific pursuits" (qtd. in Carmichael, 15–16). Male phrenologists would have preferred the support of the medical establishment and businessmen, but in the absence of such support, they readily accepted women as an alternative route to cultural diffusion.

In addition to its comparative openness, the fact that phrenology was a science that courted the amateur contributed to its appeal to women. Phrenological manuals claimed to teach the interested how to read her own character and that of others simply through self-study and practice. Far from requiring formal education or specialized equipment, the aspiring phrenologist only needed access to a few books and a willing subject to practice upon. As Alison Winter has argued regarding contested Victorian sciences more generally, "their popularity might be influenced by a perception that they were not already identified as the domain of skilled experts or elite communities" (28). Thus, although in fact only an unintended consequence of its popular appeal, phrenology emerged as a science that women could access directly, teach themselves, and apply to acquire a sense of scientific authority.

By the final two decades of the century, women were occupying positions of power and influence within the phrenological community. This is perhaps a consequence of simply being allowed to directly participate in meetings, become credentialed, and hold offices. At the May 1889 meeting of the British Phrenological Association, more than half of the members notable enough to be recognized by name in the report are women, two of whom were previously elected to serve on the executive council.[8] Annie Patenall, a professional phrenologist and lecturer in Hastings and Brighton, served as the first president of the Hasting's branch after earning her diploma from the Association.[9] Women routinely presented papers at the monthly meetings, gave public lectures, and wrote signed articles for the organization's associated journal, *The Phrenological Magazine*.[10] Jessie Allen Fowler, daughter of Lorenzo Niles and Lydia Fowler, was the most influential woman in the organization. After she journeyed to London from America with her parents in 1860, she helped to establish the English branch of the Fowlers' phrenological empire. In 1863, Lorenzo Niles Fowler founded the Fowler Institute, which Jessie helped to run (Stern 243). Earnestly interested in the structure and function of the brain, Jessie studied brain dissection at the London Medical School for Women in 1879 (238). She assumed control of *The Phrenological Magazine* in 1889, and a year after her father's death in 1896 she incorporated the magazine into the American *Phrenological Journal and Science of Health*, which was published in America but distributed from both the New York and London offices. Jessie also served as vice president of the American Institute of Phrenology and wrote several books on the science ("Jessie A. Fowler" 38).[11]

Access to a science, of course, is not in itself a challenge to the status quo, but what these women encountered in the phrenological model of mind was a physiological psychology that did not take sexual difference as the foundational premise for personal and social identity. In the eighteenth century, the dominant scientific narrative that women were "lesser" than men gave way to one in which women were distinguished from men in every possible way, from the type of organs they possessed to the size of their bones and the quality of their nerves. As Thomas Lacquer observes, when "custom became a less and less plausible justification for social relations, the battleground of gender roles shifted to nature, to biological sex. Distinct sexual anatomy was adduced to support or deny all manner of claims in a variety of specific social, economic, political, cultural, or erotic contexts" (152). This battle renewed itself with increased vigor and specificity—anatomical, physiological, and anthropological—in nineteenth-century Britain in reaction to increasing feminist agitation (Russett 10). Even without the challenge of feminism, sciences of sexual difference were perfectly suited to a separate spheres ideology rooted in a sexual division of labor, clarifying the impetus for, and sustained investment in, biological theories that naturalized women's domestic roles.

Unlike reproductive organs or secondary sexual characteristics, however, the brain was an organ that showed no apparent structural difference between the sexes. In the Victorian period, correlating structure and function in the brain *at all* was in its infancy, let alone making finer distinctions. Even the arguments about absolute size of the brain depended on a correlation between size and relative achievement, which feminists like Cobbe were quick to note might also be accounted for by women's relative lack of access to education and opportunity, a lack that could cause mental organs to atrophy. Despite their zealous attempts to enlist the brain as evidence of male intellectual superiority, because the brain could either be interpreted as a sign of women's innate inferiority or the physical record of an impoverished education, the brain became the key organ for the fight over women's rights. Given the insufficient testimony of the female brain's morphology, male scientists and anthropologists would often try to bolster their arguments with reference to women's reproductive organs as evidence of women's primary function. Pike, for instance, after acknowledging the limitations of the current state of knowledge about the brain (51), solves the issue by asserting a homologous relationship between the brain and the body, the latter being "adapted to maternity": "If there be any truth in science, the intellect of woman not only has but must have, a certain relation to her structure; and if it could be shown that there exists no difference between the male and female minds, there would be an end of Anthropology" (54, 58). It is precisely this mode of fallacious reasoning that Lydia Fowler criticizes in the opening pages of her physiology book for women:

> The remark is often sneeringly made, "that all that woman is good for, is to bear children . . . so woman can fulfil her mission, in compliance with her natural instincts, without any special education." [. . .] No one can say that the bearing of children is the *only* mission of woman . . . for the uterine organs occupy but a small space compared with the digestive and circulatory organs. A person might as well, and even more justly, say that the only mission of woman is to digest food, for this is an absolute necessity of her existence. (2)

In her response to the frequent "sneeringly made" remark that women are only fit to be mothers, Fowler pinpoints the assumption at the core of the anti-feminist argument by analogy, challenging this logic with an appeal to another form of biological evidence. She counters the "scientific" arguments against women's access to education with a form of physiological rhetoric that shifts focus from the few organs that distinguish men from women to the structurally similar organs they share. She further adduces that women are not able to give birth throughout the entirety of their lives, and thus "Nature" itself indicates that woman "has other duties to perform than those that are merely maternal" (2, 3). Lydia Fowler was herself proof that women were capable of being more than mothers: in 1863, after moving to London with her husband, she worked as a phrenologist in the family's British phrenology headquarters (Stern 182, 187). Before moving, however, she achieved prominence as the first American-born woman to receive a medical degree, supported the suffrage movement, and became the first female professor at an American medical school. Elizabeth Cady Stanton, Susan B. Anthony, and Matilda Joslyn Gage dedicate their six-volume *History of Woman's Suffrage* to Lydia Fowler and nineteen other key women in support of the movement, including Harriet Martineau, Margaret Fuller, and the above-mentioned Harriot Hunt, all of whom were also phrenological devotees. As Kristine Swenson has argued, although Lydia Fowler's accomplishments should "place her at the forefront of early women doctors in Britain and America," she has largely been erased from medical history due to her association with phrenology (110). At the time, however, it was her involvement with the Fowler phrenological empire that amplified her support of the women's movement and helped to disseminate her medical ideas, including her advocacy of women entering the profession (111). According to the *Englishwoman's Review*, Fowler's frequent lectures routinely filled the "largest halls in England," and in them she endeavored to promote medical knowledge "among her own sex in particular" ("Lydia Folger Fowler" 83).

While Lydia Fowler argued that women's reproductive organs should not determine a woman's entire social function, other female phrenologists saw the brain as an organ that could make the case for that non-determination even tighter. The morphology of the brain in phrenological theory was not easy to incorporate into extant models of sexual differentiation. For whatever reason, phrenology made no distinction between the brains of men and women in terms of the number and relative position of the mental organs. Regardless of sex, men and women possessed the same mental "parts"—in Spurzheim and Combe's model, there were twenty-one affective faculties (feelings) and fourteen intellectual faculties, for a total of thirty-five discrete mental organs. Unlike the uterus, mammary glands, or ovaries, the brain was an organ not tied to the biological aspects of reproductive function. Although phrenological literature routinely marked certain organs affectively linked to feminine character as typically being larger in women (such as "Amativeness" or "Adhesiveness"), men possessed the same organs and could even manifest them prominently without this calling into question their sexual identity. Thus, in Combe's *Elements of Phrenology*, he notes that women possess a "generally larger" organ of "Philoprogenitiveness" (love of children) but goes on to note that the capacity is also quite "large" in the brain of his personal hero and fellow Scotsman Robert Burns (65, 66).

In the case of phrenology, where the correlation between structure and function had already been (erroneously) asserted, the supposed differences between the sexes were not uniform and thus did not help to establish or codify binary distinctions. The brain's biological relationship to sex was entirely unclear, and women turned that ambiguity to their advantage. Thus, in her "Rights of Women," author and women's rights advocate Anna Wheeler uses phrenology to challenge the claim that "there is any essential difference between the sexes," and concludes that "All the researches of anatomy[] have not yet been able to prove a difference in the brain of either" (15, 33). Wheeler draws a distinction between structure and size of mental organs, emphasizing that women and men possess the same organs intended for the same functions, allowing her to point out that even if "the organs of women are generally smaller" they are nevertheless "equally fitted to the purposes for which they are intended" (33). Reasoning out from the phrenological doctrine that held that mental exercise leads to the improvement of mental function and a corresponding increase in size of the faculty, the "generally smaller" organs of women become not a cite of difference but an indictment of educational inequality. As Wheeler puts it, "nature has been quite impartial to the sexes on the score of intellect!" and it is "Difference of education alone, and the impression of objects, which surround us, under very different circumstances, produce the apparent difference in the phrenological developement of men and women" (33). She further argues that when obstacles are removed, the organs manifest equally, thereby proving that unequal education is the cause of intellectual difference rather than mental inferiority in women (34). Much like muscles that atrophy from disuse, most women's intellectual organs are weak—but because those organs are demonstrably there, structured like those of men, and entirely capable of being exercised into health and larger size, there remains no justifiable reason to educate them differently.

Wheeler's argument anticipates that of the suffragette and scientist Lydia Becker, one of the few women to infiltrate and meaningfully participate in the annual meetings of the British Association of the Advancement of Science.[12] In an address delivered at the organization's annual meeting of 1868 titled "On some supposed Differences in the Minds of Men and Women with regard to Educational Necessities," Becker offered three propositions about sex in relation to mental capability: first, that sexual difference "does not extend to mind"; second, that any apparent differences arise from the "influence of the different circumstances" that characterize the lives of men and women; and third, that in spite of environmental influences, the actual differences in mind between men and women are not any greater than differences between individuals of the same sex (484).[13] In support of her first proposition, Becker notes that the assumed correlation between physical strength and quality of mind does not hold for other species, such as plants that display distinct sexes and hermaphroditic invertebrates that cannot be said to possess "male" or "female" brains. Citing birds of prey as evidence, she further argues that the subordination of the female to the male is far from universal even among vertebrates, but she is careful to point out that superiority of physical strength is no indication of intellectual superiority (486).

Her rebuttal uses zoological facts to confound the binary distinctions of sexual difference in mind as natural or universal, a practice she sees as contradicting scientific knowledge by representing artificial categories as biological realities.

Becker's paper generated a great deal of interest and controversy, the former largely for the novelty of a women delivering a paper on a non-genteel subject. While the reception at the BAAS was not entirely hostile, the aftermath in the press was contemptuous and derisive (Parker 634–35). *The Lancet* dismissed Becker's argument as unscientific because she drew her examples from the animal kingdom rather than physiological differences between humans. Predictably, the very first piece of evidence invoked to challenge her position was the smaller absolute size of women's brains. While admitting the relativity of proportion to overall body size, the writer nevertheless asserts, "we think that men possess bigger and stronger heads for the same reason that they have stronger limbs—viz, that Nature has fitted them to do stronger work" (320). J. McGrigor Allan offered a similar critique of Becker's paper, noting that a "physiologist" cannot accept her assertion that differences in male and female minds are only "supposed" since "there are indisputably two sexes of man, who may be recognised by their appearance," a position he also substantiates with an appeal to the "relative size and conformation of the crania and brains of the sexes . . ." (202–3). Becker's position is in no way logically countered by these arguments, but it is worth noting the different type of evidence that obtain for each: for Becker, the unscientific nature of the position that sex extends to mind is due to positing a difference by analogy and without evidence; for her scientific detractors, that lack of evidence could be skirted by reasserting what was visible in lieu of what was not: stronger arms, bigger heads, and the self-evident nature of "appearance" were marshalled as bulwarks against the lack of actual knowledge about the relationship between sex and physiological psychology. Given the state of brain science at the time, Becker was right in observing that "[t]he assumed difference in the minds of the two sexes is purely hypothetical," but what she could not produce was biological evidence of mental equality.

Becker's scientific position is similar to the philosophic one taken by John Stuart Mill in *The Subjection of Women*, in which he repeatedly and emphatically claims that women's natural capabilities cannot be known because women have never not been in a state of subjection:

> I consider it presumption in anyone to pretend to decide what women are or are not, can or cannot be, by natural constitution . . . their nature cannot but have been greatly distorted and disguised; and no one can safely pronounce that if women's nature were left to choose its direction as freely as men's . . . there would be any material difference, or perhaps any difference at all, in the character and capacities which would unfold themselves. (56)

A key aim of Mill's argument was to expose the irrationality of justifying unequal treatment through an appeal to natural differences. As an environmentalist, he sought to discredit arguments by nature as merely hypothetical while foregrounding the known disadvantages and hampering conditions under which women have existed. In his third chapter, Mill directly addresses the claim that men's intellectual superiority is an effect of possessing a larger brain, offering three objections: first, that the supposed difference in size has not been established but is only inferred from a difference in stature; second, that the correspondence between the brain's physiology and psychological expression is completely unknown; and third, that magnitude is not the only measure of an organ's power. In support of this last argument, Mill briefly departs from his environmentalist position to speculate that possibly men have the advantage of size, which gives them mental stamina, while women might have the compensatory advantage of greater "cerebral circulation," granting them more rapidity of thought (64).

Although Mill hastens to add that his conjecture is "entirely hypothetical," this uncharacteristic lapse into the possibility of natural intellectual distinctions between the sexes provided Darwin with fodder for his section on the mental differences between men and women in *The Descent of Man*. In its original context, Mill's speculative detour is actually an attempt to assert intellectual equivalence in ability while accounting for possible variation in the underlying process: the brains of men and women can accomplish the same mental tasks but the physiological route might differ. Darwin, however, ignores Mill's assertion of women's possible ability to think more quickly than men and fixes on the idea of men having greater mental stamina, which he then equates with the qualities of "energy and perseverance" (ii.564 n.24). These qualities, along with "courage," form the triumvirate of innate mental characteristics that Darwin claims gave the progenitors of contemporary males the advantage over others in the struggle for possessing and retaining females. Through sexual selection and the consequent transmission of these qualities to male offspring, men had "gain[ed] the ascendency" over women, eventually becoming "more eminent in every pursuit" (ii.564). Crucial for Darwin's argument is the notion that at some point in the distant past a physiological mental distinction between the sexes was sexually selected for and increased over time. It is in entertaining the possibility of physiological difference at all that Mill's argument became vulnerable to Darwin's evolutionary perspective. There was, however, a far deeper incompatibility in their views on human psychology, since Mill saw individual human character as primarily the product of environmental circumstances whereas Darwin in *The Descent* ascribed much of human character to innate, hereditarily transmitted qualities. The foundational premise of Mill's primary argument—that women's apparent difference in character is the product of external influence—presupposes that everything that makes women seem intellectually distinct from men is actually acquired during their lifetime. Darwin, however, saw the idea of the social construction of character as fundamentally irreconcilable with his view of human evolution: "the social feelings are instinctive or innate in the lower animals; and why should they not be so in man? [. . .] The ignoring of all transmitted mental qualities will, as it seems to me, be hereafter judged as a most serious blemish in the works of Mr. Mill" (ii.98 n.5). Mill's rational argument, which in the main stressed the impossibility of determining women's natural characteristics under the constraint of sexist social institutions, could be easily dismissed by evolutionary thinkers because the sort of evidence their paradigms privileged were entirely different: the inherent versus the experientially acquired.

Unlike Becker and Mill, women phrenologists could challenge an argument based in natural capacity on its own terms by granting the primacy of the brain's materiality but interpreting it in a different way. Thus, in response to the question, "'Will it be worth while for women to enter the lists with their brothers in competitive examinations?'" Jessie Fowler answers, "female ambition in matters of collegiate competition—as in every intellectual sphere—must of course be based upon mental capacity; where that capacity exists . . . let women be tested with their brothers" ("Spinsters" 110). And, according to her, the proof of this capacity was already available in the comparative study of the brain. She writes: "Phrenology clearly points out that the brain power of woman is equal to that of man," a position supported by "the evidence given by the most eminent phrenologists and physiologists" (109). The notion of a materially demonstrable capacity offered an entirely new basis for claiming political and social parity by transforming the hypothetical into the empirical. It was no longer, as Mill and Becker had claimed, *not* knowable what women could accomplish if given the chance—their equal capacity was obvious, clearly delineated in the structure of the brain. Thus, providing women with educational and professional opportunities could not possibly be a waste of resources, for the structure of the brain itself was proof of possibility, and to not provide women with such opportunities was a waste of brain power. As one feminist phrenologist put it in an article titled "The 'Coming Out' of Woman":

> Women have discovered that they possess faculties that, when trained, enable them to do most of the work that men are accustomed to do. The fact of organization seems heretofore to have been disregarded, so that fully 40 per cent of the intellectual organs of the brains of the average woman has been in an inactive state. We need not say that this has much the character of a wrong, since Nature intends that with the possession of a faculty its use should be associated. (109)

The proposition of women's demonstrable but untapped mental power thus emerged as a violation of the self-evident ethics of a natural system. The voice that has the final say on women's proper place in the world is Nature itself, with an authority that supersedes law, custom, and tradition. This use of biological essentialism presents the inverse of the equally determinist argument proposed by male anthropological and evolutionary thinkers—that is, that social inequity follows from and reflects natural differences in ability. Feminist phrenologists, while still retaining the authority of nature, assert the opposite: the current state of society does not reflect the biological fact of women's mental ability, rendering the sociopolitical reality of the day both irrational and unnatural. The physiological capabilities of the woman's brain reveal mental resources that have been neglected and thus underutilized, and it is the very fact of its being there—the material, empirical, visible evidence of anatomical structure—that justifies the claim to equal rights.

The Phrenological Journal and Science of Health made more implicit arguments for women's equal mental capacity on physiological grounds.

For example, in the series "English Men and Women of Note," the journal published portraits of public figures accompanied by a phrenological reading that identified and described the mental capacities confirmed by their biographies.[14] Such illustrated biographical narratives had long been a staple of phrenological publications, both as a way to provide evidence of the validity of the science and as a way to capitalize on the public's curiosity about the psychology of famous (and infamous) individuals. In this case, however, each installment includes both a man and a woman remarkable for their intellectual accomplishments. For instance, one issue pairs Margaret Oliphant with Arthur Conan Doyle, pointing out how the former's ability to write realistic novels is owing to her possession of a forehead "well rounded out in the upper front crest," while the latter's talent for detective fiction is attributable to "the breadth of the head, especially the frontal lobe" (159, 160). Another pairs Sir Isaac Pitman, inventor of phonetic shorthand, with the "The Famous Lady Journalist" Emily Crawford. As in Oliphant's case, the article stresses the underlying mental capacities that equip Crawford for her profession, including a "large Individuality, Eventuality, Comparison, Locality and Intuition" that enable her to easily remember statistics, locations, and people (113–14). By pairing professional men and women in this way, the series makes an implicit argument about the mental equality of the sexes—not in terms of similar ability (as every person is, phrenologically speaking, entirely unique), but in terms of possessing mental capabilities that qualify them for specific occupations. The rhetorical gambit is in positing an equivalence of physiological causality: just as Pitman's brain predisposed him for invention, Crawford's fitted her for journalism; just as Doyle's organization suited him for plotting detective stories, Oliphant's faculties prepared her for writing realist fiction. The attention to the portraits that precede the biographical description further underscore the demonstrable nature of mental aptitude, which the "photograph indicates," or which becomes clear "when looking at the excellent portrait before us" ("Mrs. Oliphant" 159). The emphasized aspect of the evidence presented is visual, material, and above all, biological. Rather than relying on the hypothetical supposition that women's intellectual capabilities might flourish under altered social conditions, the phrenological argument stresses the immediacy of ocular proof and the self-evident nature of physical signs. Thus, irrespective of their life experiences, women with or without the particular advantages of those few who have managed to distinguish themselves might nevertheless see themselves as already biologically and physiologically equipped for those professional occupations some of their sex had already obtained.

Yet perhaps the most dramatic shift occasioned by phrenology in relation to gender identity was not about the innate capacities of women as a class but about the variety of personal identities available to women as individuals. Because phrenological discourse never delineated a separate or distinct method for assessing the character of women, it had the collateral consequence of making apparent the incommensurate relationship between women's biologically inscribed mental individuality and the limited social roles available to them. Take, for instance, the format of phrenological delineation booklets that were produced, marked, and sold by practical phrenologists: the first few pages of these booklets typically contain a list or table of the mental organs along with their possible manifestations in size to be marked by the examining phrenologist (Figure 1.2).

The remaining pages explain at length how these various faculties display themselves in the client's character, following every category with instructions on how either to restrain or cultivate the manifestation of each. These booklets were sold to clients who wished to further investigate, through self-examination and introspection, their own character after a reading. This ingenious commodity had the rare ability to be one size fits all and bespoke at the same time: every client received the same book, but every book told a different story of self.

Yet of all the nearly infinite narratives of physiological development these booklets might tell, none pointed to the limited domestic roles of the private sphere. A woman with the well-developed organ of "Philoprogenitiveness" would also have every other organ marked, revealing her to be innately more than her domestic function. In booklets that also went on to recommend professions based on development, the many possibilities might include "Naturalist," "Lawyer," "Engraver," "Contractor," "Grocer," or "Detective," but they did not include "domestic helpmate." This stands to reason: motivated by profit, practical phrenologists simply did not go to the trouble or expense to produce separate booklets for women, but they would take their money all the same.

As Roger Cooter has shown, phrenology's emphasis on innate individuality was key to its ideological significance in naturalizing social relations under industrial capitalism. While it technically worked against the interest of the working class by rationalizing the hierarchal division of

V.

For an Explanation of this Table the Reader is referred to the pages immediately following (viz. vii., viii., and ix.).

Conditions	Over Active	7 Very Large	6 Large	5 Full	4 Ave'ge	3 Moderate	2 Small	1 Very Small	Inactive	Cultivate	Restrain
Organic Quality	69	70	70	70	70	71	71	72	..	72	72
Health	..	73	73	73	73	73	74	74	..	74	..
Vital Temperament	..	80	82	82	82	82	83	..	..	83	83
Lymphatic Form	..	84	85	86	86	86	86	86	..	86	86
Breathing Power	..	88	88	88	88	89	89	89	89	89	..
Circulatory Power	..	91	91	91	91	91	91	91	..	92	..
Digestive Power	..	94	94	94	94	94	94	..	94	95	..
Motive Temperament	..	96	96	97	97	97	97	..	..	98	98
Bilious Temperament	..	99	99	100	100	100	100	..	..	100	..
The Kidneys	..	101	101	101	102	102	102	102	..	102	..
Mental Temperament	..	104	105	105	105	105	106	106	..	106	106
Evenly Balanced do.	..	107	107	107	107	107	107	..	..	107	..
Activity	..	108	108	109	109	109	109	..	..	110	110
Excitability	..	111	111	111	111	112	112	112	..	112	112
Size of Brain	..	118	119	119	119	119	120	120	..	120	..
The Social Group of Faculties	..	120	120	121	121	121	121	121	..	121	121
Amativeness	123	123	123	123	126	126	127	127	127	127	127
Conjugality	128	129	129	130	130	130	130	130	130	130	131
Parental Love	131	132	132	133	133	133	133	133	133	133	134
Friendship	135	136	137	137	138	138	138	138	138	138	139
Inhabitiveness	139	139	140	141	141	141	141	141	141	142	142
Continuity	142	143	143	143	144	144	144	144	144	144	144
Selfish Propensities	..	145	145	146	146	146	146	..	..	146	147
Vitativeness	147	147	148	148	148	148	148	148	149	149	149
Combativeness	151	151	151	152	152	152	153	153	153	153	153
Destructiveness	156	156	156	156	156	157	157	157	157	157	157

Figure 1.2. Table of phrenological organs from R.B.D. Wells's *A New Illustrated Hand-book of Phrenology, Physiology, and Physiognomy*, London, H. Vickers, [1860?].

vi.

The Printed Figures in the square marked by the Examiner indicates the page in the book on which the organ is described.

Conditions.	Over Active	7 Very Large.	6 Large.	5 Full.	4 Ave'ge	3 Moderate.	2 Small.	1 Very Small.	Inactive.	Cultivate.	Restrain
Alimentiveness	159	160	160	160	160	160	160	160	160	160	161
Bibativeness	161	162	162	162	162	163	163	..	..	163	163
Acquisitiveness	164	165	166	166	166	166	166	166	166	167	167
Secretiveness	167	168	168	169	169	169	170	170	170	170	170
Cautiousness	171	172	173	173	173	173	173	174	174	174	174
Approbativeness	176	176	176	177	177	177	177	177	177	178	178
Self-Esteem	..	181	181	182	182	182	182	182	..	182	182
Firmness	..	183	184	184	184	184	185	185	..	185	185
Moral Sentiments	..	186	187	187	187	187	187	187	..	187	187
Conscientiousness	..	188	189	190	190	190	190	190	..	190	190
Hope	.	191	192	193	193	193	193	193	..	193	193
Spirituality	..	195	196	196	196	196	196	196	..	196	196
Veneration	..	198	199	199	199	199	199	199	..	199	199
Benevolence	..	200	201	202	202	202	202	202	..	202	202
Self Perfecting Group	..	203	203	204	204	204	204	204	..	205	205
Constructiveness	..	205	205	206	206	206	207	207	..	207	207
Ideality	..	207	209	209	209	209	209	209	..	209	209
Sublimity	..	212	212	212	212	212	213	..	..	213	213
Imitation	..	213	214	214	214	214	215	215	..	215	215
Mirthfulness	..	216	216	216	217	217	217	217	..	217	217
The Reflective and Perceptive Faculties	..	222	222	222	222	223	223	223	..	223	223
The Perceptive Faculties	..	223	224	225	225	225	226	226	..	226	226
Individuality	..	228	228	228	228	229	229	229	..	229	229
Form	..	230	230	230	231	231	231	231	..	231	..
Size	..	231	232	232	233	233	233	233	..	233	233

labor as "natural," it nevertheless appealed to this very class by promising happiness, advancement, and success through self-knowledge. A young man desirous of upward mobility could use phrenology to determine the best trade for his inherent talents, discover what innate qualities he might cultivate through self-discipline, and reconcile himself to his position in the world by finding it of a piece with a natural social order. As Cooter puts it: "phrenology provided artisans with what in retrospect can be seen as a rationale for accepting more individualist-orientated forms of self-improvement . . ." (*Cultural Meaning* 172). Women received the same information from a phrenological reading as men: an evaluation of innate talents and proclivities and an injunction to improve through self-culture, introspection, and faculty exercise. Yet, insofar as the reconciliation to capitalist society Cooter describes was premised upon the division of *skilled* labor that supposedly followed from one's specific physiological predispositions, women were actually barred from experiencing the same sense of peace regarding their place in society. Because of the enormous distance between the degree of socioeconomic, political, and legal freedom granted to working men and their female contemporaries (of every class), the science that operated as a hegemonic force for the former had the unintended effect of providing a rationale of liberation for the latter—paradoxically, precisely because it was never "for" them in the first place.

The biological aspect of phrenology allowed it to carry a different ideological valence for women than for men, in part because it made discovering and cultivating one's innate individuality a moral imperative. Part of the science's widespread appeal lay in its claim that everyone, regardless of class, rank, or wealth, possessed a completely unique personality. The mere fact that there were thirty-five distinct organs that could each be measured in a range of manifestations ("small," "moderate," "full," "large," etc.) meant that there were more possible combinations of organs and sizes than there were humans on earth, then or now.[15] Thus, long before the discovery of DNA, the apparent physiological fact of innate individuality revealed and justified to those interested in the science an object for study centered in the self. As Annie Patenall explains in *The Phrenological Magazine*, "we are individually unlike every other being in the universe. . . . What we have to do therefore to arrive at personal perfection is, to learn as much as we can of our own nature and destiny . . ." (228–29). The phrenological motto "Know Thyself," emblazoned on busts, advertisements, and charts, implied that individuals did not know themselves already, and that it was thus one's duty to discover and cultivate that true identity once known. The phrenological injunction to obtain self-knowledge, however, spoke differently to women than it did to men because their delineations reflected an individual identity that, while unique, was also not able to be fully recognized or nurtured by external circumstances. As Jessie Fowler explained to the British Phrenological Society after being recognized for her efforts as an itinerant female phrenologist lecturing in Australia:

> We have only just begun to understand ourselves, and yet what subject is there that is more interesting. You talk to persons about themselves, they get interested, and you fasten the thing down in a very practical kind of way. [. . .] I was grateful for the words of the Chairman in regard to my being a woman, because I have realized the difficulty that I have had to contend with ever since I began this work. [. . .] If the community were composed wholly of men, then it would be wholly man's work; but where the community is divided, where there are so many women who do not understand themselves, it seemed to me, as I had the spirit given to me, that I should be wrong to myself, and do a wrong to my Creator, if I did not answer that voice, and so I have done what I could. ("British Phrenological Association" 184–85)

This forthright account reveals the practical advantage of offering clients information on the most universally interesting subject: their own character. An appeal to human egotism was the hook, but the self that was revealed might contain within it the justification for its own expanded range of agency. Allying biological revelation with spiritual duty, one woman's lecture tour assumes the character of mission work, yet one in which the message is women's new knowledge of themselves as individuals.

The idea of an innate individuality in women necessarily challenged the "scientific" argument from nature that women's identity was essentially based in maternal function. If the justification for women's social role was premised upon reproductive biology, what of the equally biological nature of the brain? The mind's materiality became the site of a deep identity, hidden from conscious awareness and obscured by the prevalent social ideology that women should find their primary sense of self in relation to men. As one woman writes in *The Phrenological Mag-*

azine, "one of woman's greatest inherent dangers [is] a tendency to seek her own individuality in that of the other sex. . . . We are individuals. We are responsible creatures, just as much as men . . ." ("Individuality of Woman" 223, 224). She goes on to identify this tendency in women as a value encouraged by conduct manuals, celebrated in marriage, and promoted by clergymen even among unmarried women. She urges women to "live their *own* fresh natural lives, instead of tamely echoing those of others" (224). In this formulation, what is "natural" to women is an inherent identity that existed before the interference of social relations. Thus, to center one's life on promoting the well-being of men emerges as an unnatural act of self-erasure.

One of the most practical consequences of recognizing women's biological individuality as something fit for more than marriage and motherhood was the acknowledgment of their mental aptitude for professional roles. For instance, a book by the husband-and-wife team of Stackpool and Geelossapuss O'Dell includes numerous case studies describing female heads. As the leading London phrenologist in the 1890s, Stackpool received an average of 15,000 customers per year, and the great variety of female skulls suggested to him a disjunction between mental capability and professional opportunity. He explains that if a girl has a head that "denotes all the abilities required for a successful architect, farmer, lawyer, or doctor," she is seldom encouraged to pursue these fields because the "conventionalities and social arrangements are such that the exercise of their abilities is more limited than that of boys" (100, 101). He further laments that "we come across a number of girls who possess literary faculties, some of whom might be well suited for the production of good stories or journalistic work . . . [but] if the phrenologist mention, before the girl, the possibilities of her organisation in this direction, the mother will consider her visit more injurious than beneficial, as it may put 'notions' into the child's head" (101). Despite the unwelcome reception of such information, O'Dell explains that he persisted in informing clients of their girls' innate capabilities: telling a mother that her daughter's "mental faculties strongly indicated the intelligence requisite for the pulpit" (101), informing a grandmother that her granddaughter had the "clear and intellectual grasp" that would make her an excellent teacher (111–12), and explaining to an aunt that her niece should be trained in landscape gardening and seek a position in the Kew Botanical Gardens (106).

The power of physiological self-knowledge to catalyze personal development among women is charmingly rendered in Elsie Cassell Smith's serialized novel *The Amateur Phrenological Club (Its Sayings and Doings)*. The plot details the experiences of a group of "intellectual women" who form a society to study and discuss the science. The book includes occasional adventures—for instance, when the group vacations at a seaside resort and saves a young woman from becoming engaged to a man with a skull displaying a horrifying basilar region. The central focus of the narrative, however, is on the personal development of the women based on a mutual recognition of their inherent talents. The members eventually adopt the practice of referring to one another by epithets based on their psychological aptitudes, such as "the Artist," "the Philosopher," and "the Executor." By the novel's conclusion, the promise of every title is fulfilled: the Artist becomes a famous portrait painter, the Philosopher receives a degree from the Phrenological Institute and finds success as a traveling lecturer, and the Executor serves as the Philosopher's intrepid business manager.[16] Throughout, the narrator repeatedly dramatizes the individual and interpersonal dynamics of confidence-building based on the biological warrant of capability. When the Philosopher contemplates taking a course of study in Paris, she reports, "Our gifted artist friend had been the recipient of much encouragement and commendation from us all, because of her perfect adaptability, phrenologically, for her chosen profession." Similarly, the Philosopher's "excellent adaptability to the noble work of reform" is foretold by the "fineness and firmness of her organic quality" (230).[17] In this way, *The Amateur Phrenological Club* presents physiological evidence as proof of possibility, using materialism to both justify women's personal ambition while at the same time cultivating, through mutual encouragement, the conditions that bolster professional development. In the context of an all-women's organization in which members interpret their own physiologies, biological determinism assumes a character far different from that in anthropological or racial contexts. This is because what essentialism projects in this case does not reify existing hierarchies, but rather challenges them: while they may begin as amateur phrenologists, they end as independent and successful professional women.

This essentialist reasoning had its nonfictional counterpart in phrenological publications that cited women's intellectual accomplishments as evidence of their innate intellectual abilities. In the last two decades of the nineteenth century, *The Phrenological Magazine* routinely published articles advocating for the expansion of educational and professional opportunities for women by foregrounding the success of those who had distinguished themselves intellectually.[18] An essay titled "Female Astronomers," for instance, counters the claim that women are "not qualified by her intellectual endowments to accompany [men] in the higher fields of art, literature, and science" by offering a series of biographical vignettes to show women's capability of "studying with success that most abstruse and difficult of the sciences—astronomy . . ." (419). The author details the scientific careers of twelve women, beginning with Hypatia, ending with Mary Somerville, and including William Herschel's much overshadowed sister, Caroline Lucretia. This is the kind of feminist work that aims at historical recovery—an attempt to reinstall and recognize women's otherwise overlooked, forgotten, or undervalued intellectual contributions. It also, however, reifies an essentialist argument that women's intellectual capabilities are equal to those of men by foregrounding the accomplishments of women who had the benefit of educational opportunity.

The argument that the physiological structure of women's brains had a one-to-one correspondence with impressive intellectual capabilities was also emphatically asserted in articles presenting phrenological analyses of professional women's photographs. Under Jessie Fowler's editorship, *The Phrenological Journal and Science of Health*, which was distributed in both the United States and England, began running such articles as a regular feature, increasing women's professional visibility both figuratively and literally. The articles typically begin by underscoring the most impressive mental faculties displayed in the photograph, followed with a biography that confirms the existence of those innate aptitudes. A representative article analyzing a photograph of Mary Lyon Dame Hall, President of Sorosis (the first professional woman's club in the United States), points to how her "brain appears to be a very active one, and particularly so in the executive, intellectual, moral, and social regions," the promise of which is fulfilled by her having become a prominent figure among [Sorosis] members on account of her intellectuality . . ." ("Mrs. Mary Lyon" 14, 18). A similar article analyzing the profile of Reverend Antoinette Brown Blackwell, the first woman ordained as a Protestant minister in the States, draws attention to her "Vital, Motive and Mental Temperaments," confirming that "from a physiological standpoint, her mentality has a good foundation to work upon. She could not preach dyspeptic or gouty sermons if she tried, and hers, by the way, is the kind of healthy life we need in the pulpits of to-day" (Jessie Fowler, "Rev. Antoinette Blackwell" 118). As I have argued elsewhere, the organizational structure of phrenological analyses of illustrations, which subordinated biographical narrative to pictorial and psychological analysis, popularized the idea that character had a biological component that caused human behavior, rather than it being entirely the effect of environmental circumstances.[19] For the majority of these texts, particularly in the earlier period of phrenology's popularization, the rhetorical purpose is simply to establish the science's validity. These illustrations were almost always of famous or infamous men whose biographies were already well known, and the work of the text was to reveal the ways in which the shape of his skull foreshadowed the behavior that established his notoriety. In articles analyzing the heads of notable women, however, the authors simply assume the legitimacy of the science and focus more on extending the argument of biological causation to a larger social context. The brain's materiality becomes the primary qualification for women's occupations, thereby detaching professional identity from experience that can only be gained through opportunity. In this formulation, woman's innate identity, encoded in the brain's structure, attests to her abilities and justifies her right to pursue the occupation for which nature has suited her. Insofar as this biologically based reasoning had a predictive quality (e.g., Antoinette Blackwell is physiologically *incapable* of delivering a poor sermon), it provides an anticipatory counterargument to the objection that, if given the opportunity, women might not be able to perform well in a professional capacity. Society had nothing to lose and everything to gain by welcoming women into the professional workforce because women who followed the dictates of their mental natures could not help but succeed.

Such articles also leveraged the language of self-culture that was so integral to phrenological doctrine in order to contradict a common argument opposing women's intellectual development on medical grounds. For instance, an article on Caroline Hazard (fifth president of Wellesley College) comments that "Instead of being dyspeptic, gouty, or nervous," Hazard "looks as though her cap and gown had cost her many hours of pleasurable study, and instead of being ready to die of nervous prostration we could hardly select a better example of health, and her influence will be accordingly beneficial" (J[essie] F[owler], "Miss Caroline Hazard" 384). The defiant tone of this visual appraisal defensively responds to the medical assumption that the mental strain of educational and occupational activities traditionally performed by men had deleterious effects on women's health. As Cynthia Eagle Russett has shown, building on the principle of the conservation of energy, physicians and psychologists from the 1860s onward championed an "economy of the body" (113) in which the various biological functions competed for resources (105–19). In this model, the considerable expenditure of energy required for the regulation of women's reproductive system meant they would have little left over for serious thought. Far from damaging her physical condition, however, Hazard is cast as flourishing in the wake of her educational accomplishments. In this formulation, the consequence of mental exertion is psychological health and happiness precisely because the internal capabilities of the individual are gratified by external circumstances. This follows the logic of Combe's central argument in *The Constitution of Man in Relation to External Objects*: that one of life's greatest imperatives is discovering the constitution of one's innate character and bringing environmental circumstances into alignment with those physiological dispositions. Only when experience of the outside world gratifies the unique organization of a particular mind will a person enjoy "a fountain of *moral and intellectual happiness*, which is the appropriate reward of that obedience" (19). Keying into this mainstay of phrenological logic but applying it to the domain of women's intellectual labor, the bourgeois value of self-culture becomes, in a different register, a physiological argument for the higher education of women. The robust health, vigor, and contentment putatively evident in the photographs of Hazard and the numerous other women profiled in the magazine ratifies the existence of mental organs not only capable of intellectual labor but in need of such labor for the maintenance of psychological and physical health.[20]

Throughout such articles, the authors' use of progressive rhetoric makes clear that the women analyzed are not aberrations of their sex but, rather, representative of intellectual classes that have hitherto been denied educational access and professional opportunity. The essays also draw attention to how such women, through their physiological presentation and occupational accomplishments, contradict stereotypical assumptions about women's innate capabilities. An article on Mrs. James Fairman, an early graduate from New York University's Women's Law Class, states explicitly its aim to "encourage more members of her sex to take up the same course that she has been graduated in" (45), going on to name the sympathetic professor and teaching assistants like-minded women should seek out in following her example (46).

ADMIRAL CERVERA.

Judging by the outlines of this head we find it is well developed in the lateral portions, as well as presenting a good forehead. The side-head indicates reserve, diplomacy, tact; the forehead planning talent, thoughtfulness, and organizing power. He is certainly a clever man, and has played his game well; but the last move gave more opportunity to the American fleet to dis-

which women have been engaged only comparatively for a short period when comparing this profession with medicine; therefore it is with pleasure that we present to our readers a lady who has taken up the study at the Women's Law Class of the University of New York City, and we do this for several reasons: To encourage more members of her sex to take up the same course that she has been graduated in, with the object of private or personal benefit;

MRS. JAMES FAIRMAN.

tinguish themselves, while if he had remained in the harbor of Santiago he could have given more trouble to Sampson and Schley.

MRS. JAMES FAIRMAN.

In our Phrenological Sketches of Women Engaged in Medicine, Philosophy, Teaching, or Business, which now reaches a score, we have come to a very interesting and rather novel department of work, namely, Law, in

because she is one of those noble workers who is foremost in the ranks of energetic women, and because, with all her club work and executive duties, she is an example of thorough womanliness and one devoted to philanthropy. It is in the latter cause that she perhaps takes the greatest delight, and therefore we will not leave it unmentioned.

Law in itself was supposed to be out of the province of woman altogether, although the life of one of the first and greatest law-givers (who was no less a

Figure 1.3. "Mrs. James Fairman," *The Phrenological Journal and Science of Health*, vol. 106, no. 2, 1898, pp. 45–47.

Fairman's "very vigorous brain," represented by her broad brow, contradicts the popular view that "Law in itself was supposed to be out of the province of woman altogether" because woman "is supposed to have no logical talent at all" (45, 46). The photograph accompanying Fairman's delineation presents her in cap and gown, visually unifying her inherent capabilities and the proof of their existence through accomplishment (Figure 1.3). Similarly, a profile of "Mrs. J. Ellen Foster, LL.B." provides proof of women's suitability for the practice of law due to the "positiveness in her face and head," which attests to her formidable "energy and practical insight." This is borne out by her successful career as a lawyer and election to the office of President of the Women's Republican Association (T.A.F. 148). What begins as a physiological and biographical sketch, however, soon gives way to the real rhetorical aim of the piece: to encourage women who attend college to follow in Foster's footsteps and become practicing lawyers themselves (149). While the women analyzed in such articles are important as individuals who have achieved professional success in fields rarely open to their sex, their greater significance is as biological representatives of the intellectual potential of women's minds. Ultimately, the profiles offer striking examples of certain mental "types"—doctors, preachers, lawyers, scientists—whose latent abilities have long remained untapped in women, but are nevertheless there. Wielded in the service of expanding opportunities for women, biological essentialism in this context offers proof of equal capability, and makes that case not just against patriarchal discourse but to women themselves.

By the second half of the nineteenth century, phrenology had lost its battle for legitimacy within established scientific circles, whereas craniometry gained and maintained its influence among the key exponents of biological determinism until the beginning of the twentieth century. Yet it was during craniometry's reign that phrenology became even more imbricated in everyday social life due to its accessibility and apparent practical applicability. While the (equally erroneous) numerical evidence of women's intellectual inferiority filtered down from many of the most renown scientists of the day, a second theater emerged within culture, using popular science to contradict these claims from below. It is, of course, impossible to know the degree to which this more subversive form of phrenology, mostly promulgated by women, held sway over the minds of the public. Nonetheless, it was one strand among many contributing to the growth of the women's movement in the nineteenth century, and it was certainly the most concerted use of biological determinism for feminist purposes in the Victorian age. What is ultimately important about recovering this work is not what it tells us about phrenology but what it reveals about the forms of scientific evidence and authority available to women at that time and how they managed to strategically deploy it for their own social and political aims.

While feminist phrenology lost the battle for the brain within the scientific establishment, it was nevertheless a small group of women scientists at the University College of London who managed to put an end to craniometry by analyzing independently collected data through an even more rigorous form of numerical analysis. In 1901, Alice Lee published a paper based on her thesis in the *Philosophical Transactions of the Royal Society* that argued that no correlation existed between intelligence and skull size. Lee built on the work of a "Miss C.D. Fawcett," whose research at the university had shown that the differences in determining mean skull volume (using sand, shot, or seed) between two different experimenters on the same collection of skulls differed by as much as forty cubic centimeters. As an alternative to the observational bias discovered by Fawcett, Lee advanced a formula based on an objective "mathematical theory of correlation" (228). Lee then assessed the capacity of thirty-five male anatomists (256), twenty-five male teachers at the University College (257), and thirty female students at Bedford College (258). Her results revealed that a number of the female students had larger skull capacities than the anatomists; thus, to continue to insist on a correlation between absolute size and intelligence would also mean accepting that there were young women at Bedford who were far more intelligent than some of the most distinguished gentlemen of science. She concluded that since the correlation did not hold for individuals, there was no "solid quantitative" basis for the claim that there was a distinctive correlation showing difference between the sexes. Further, she observed that skull capacity was probably "not correlated with racial ability" either (259). The next year, Lee co-authored a paper with Marie Lewenz and Karl Pearson that mathematically trounced the many objections to Lee's first paper and showed the supposed correlation between absolute size and intelligence was less supportable still. As Elizabeth Fee has pointed out, Pearson's position at the University College of London, standing in the field of statistics, and formidable character played an important role in supporting the work of these women and eventually turning the tide against craniometry, which by 1906 was entirely defunct

(431–32). Pearson supported women's rights and demonstrated that commitment in his mentorship of his female students. It was, however, Lee's groundbreaking research that provided the foundation for dismantling the accumulated force of the craniological evidence cited by such men as Darwin, Broca, Vogt, and Romanes. As Pearson clarifies in his note introducing Lee's paper in the Transactions, his "task [had] been that of an editor," merely (225).

The contribution of Lee, Fawcett, and Lewenz to brain science is now a footnote in history, but they were nevertheless instrumental in undermining a sexist and racist methodology that had provided the central scientific pillar of antifeminism for half a century. While their approach differed markedly from that of the female phrenologists who proceeded them, their story offers a fitting coda for this chapter due to the similarities of interest, social position, and strategy. Like these pioneering women, feminist phrenologists sought admission into a scientific community, however defined. They also strove to reclaim the discourse surrounding the brain from the sexist biological determinism of evolutionary psychology and anthropology. Most crucially, however, they shared with these biometricians a conviction that there was something to be gained from leveraging the rhetorical force of quantitative, empirical, and biological evidence for subversive ends. They were not alone in this conviction, as the following chapters will show.

Chapter 2

Of Two Minds

Charlotte and Anne Brontë's Use of Innate Psychology

In an article "On the Application of Phrenology in the Formation of Marriages," the Secretary of the Dundee Mechanics' Society, Alex Smart, argues that the organic nature of the brain should be of the "utmost importance in leading to the choice of a suitable partner." To support this position, he presents a series of stories about unhappy women who had the misfortune of marrying men with heads incompatible with or inferior to their own. One of the more affecting cases runs as follows:

> a young woman in whom the domestic and moral faculties were strong, and whose intellect was considerable, married a man about her own age, with great force of character, resulting from a large head, and with large animal and intellectual, but deficient moral, organs . . . by degrees, the husband acquired dissipated habits, and neglected his domestic duties. His wife used every endeavor, by mildness and persuasion, to reclaim him, but, from his deficiency of the moral faculties, without effect . . . the younger members of the family inherit the strong animal faculties and deficient morality of the father. The mother confesses she has had little moral enjoyment, and she feels that the remaining portion of her life is to be embittered by the profligacy of her children and the unfeeling indifference of her husband. (470)

Smart's decision to concentrate on the misery of wives rather than husbands makes sense: because a woman's legal personhood was subsumed into that of her husband's after marriage, the stakes of marrying someone with a poor character were far higher for women than for men. Whereas a dissatisfied husband could freely leave the home, access his own money, and retain custody of his children, a married woman had none of these resources or protections. The writer's insistence that this unfortunate wife "used every endeavor, by mildness and persuasion, to reclaim him" is similarly strategic; after all, if a wife *could* exercise a domesticating influence on her husband, there would be no need for a phrenological assessment of his character before marriage as the onus of correcting him would rest on her.

Anecdotes like Smart's were exceedingly common in phrenological texts, and with good reason: by using short narratives of domestic infelicity, popularizers could make a straightforward case for one of the most practical applications of the science. In doing so, however, they necessarily contradicted the pervasive Victorian ideology of women's moral influence—one of the few privileges that compensated women for a wife's subordination in marriage. As Sarah Stickney Ellis writes in her popular conduct manual *The Daughters of England*, "women, in their condition in life, must be content to be inferior to men; but as their inferiority consists chiefly in their want of power, this deficiency is abundantly made up to them by their capability of exercising influence . . ." (19). As Nancy Armstrong has shown, conduct manuals and novels that touted a women's regulatory power within the domestic sphere helped to create a new feminine ideal that, in turn, promoted the ascendency of the middle class. Armstrong's argument reinstalled into literary history women's centrality to the hegemonic forces that subtend modern culture. However, a bid for power that relies on claiming the domestic as a woman's special province necessarily depends on a separate spheres ideology that grounded itself in the biological distinctions between the sexes in both body and mind.

While feminine influence accorded women with a degree of power, it nevertheless also helped to reinforce and legitimate the relational dynamics that undergird the subjugation of women to men. In this model, a man's immoral behavior reflects his wife's domestic failings, making her culpable for his lack of virtue. Further, it promoted the idea that a woman's psychological existence should concentrate itself on the well-being of her husband, making her value dependent on her service to him. Certainly, the gendered division between the public and private was the dominant marriage model in the Victorian period, and many women authors did find, in the ways Armstrong describes, "polite" routes to socioeconomic power through domestic influence (97). There were, however, other ways that women writers made arguments for increased authority without relying on a model of sexual complementarity. One such route was by appealing to the authority of nature, particularly when it contradicted the logic of the dominant ideology regarding women's rights within marriage.

This chapter seeks to show how Anne Brontë—in ways far different from her sister, Charlotte—used biological essentialism to radically challenge the efficacy of domestic influence in her novel *The Tenant of Wildfell Hall*. Although this novel reflects the idea that environmental circumstances can have a mediating effect on character formation, its more emphatic insistence on the naturally encoded limits to psychological change underwrites and legitimizes Helen Huntingdon's break with ideological, social, and legal codes. Anne Brontë's incorporation of phrenological language and physiological traits is not a simple reflection of a popular psychology of the day, but rather a way to think through the premises that underwrite social and cultural expectations about a woman's duties as a wife and mother. Ultimately, it is Helen's realization that she is married to a man who is congenitally predisposed to abusive behavior that makes clear to her the contradictions inherent in ideological notions about feminine domestic responsibility.

In this respect, Anne Brontë's use of the science is opposed to that of her sister Charlotte, the author most frequently linked to phrenological discourse in literary scholarship.[1] This attention is due in large part to Sally Shuttleworth's landmark study *Charlotte Brontë and Victorian Psychology*, which aligns Charlotte Brontë's use of phrenology with the dominant scholarly interpretation of the science's role in Britain, underscoring how it naturalized the dominant ideology of the middle class, promoted the virtue of self-discipline through an internalization of surveillance, and contributed to creating the illusion of a self-regulating capitalistic society.[2] As explained in the previous chapter, however, this pervasive science was available for a wide range of social uses, whether reactionary, reformist, or radical. Although this chapter places Anne and Charlotte Brontë's use of physiological psychology in relation to each other, it is not in order to comparatively judge them on an evaluative scale of ideological correctness. The radical feminist implications of

Charlotte Brontë's novels have long been established, as well as how that feminism lies uneasily with issues of colonialism and race. This chapter, however, seeks to clarify the ways in which Charlotte's use of an innatist psychology differs in important ways from its use in Anne's work, even while it is invoked in strikingly similar contexts. Where and how they diverge at these common sites—including mate selection, education, and heredity—brings into focus the more subversive uses of biological materialism that have been obscured by its more reformist or reactionary applications.

One reason this application of the science differs so dramatically between the two sisters is that Anne's work considers the social implications of a biological basis of psychology *for women*. The relationship between the role biology plays in determining individual psychology and a married woman's rights is a major concern of the novel, and it is a concern that has little to do with upward mobility or self-discipline. In this respect, her use of physiological psychology is much more in line with the subversive appropriation of science by the women phrenologists discussed in the last chapter.[3] Yet, despite its significance in her work, almost nothing has been said about Anne Brontë's extensive engagement with physiological theories of character.[4] Shifting focus from Charlotte to Anne in a consideration of physiological science reveals an entirely different perspective on the social implications of an innatist psychology: whereas Charlotte's perspective is of a piece with mainstream Victorian values, Anne's work engages with the underlying implications of natural limitations—particularly in relation to a culture that idealizes the power of women's nurture.

The Brontës' familiarity with the science is evident in their frequent and specific use of phrenological language, but the Haworth context offers evidence of the many ways in which the science permeated even a rural community. One conduit was the Keighley Mechanics' Institute, which Patrick Brontë joined in 1833, and from which the Brontës were known to have borrowed books (Dewhirst 35). As Shuttleworth has shown, this particular institute acquired a *Manual of Phrenology* in 1837, and a year later a phrenological reference appears in one of Charlotte's early stories (*Charlotte Brontë* 57). From the 1820s on, mechanics' institutes increasingly associated themselves with phrenological doctrine, both to increase membership and because the ideology of phrenology aligned perfectly with the institutes' original aims of self-improvement through adult education.[5] In service of this affiliation, institutes carried phrenological material in their libraries, sponsored phrenological lectures, and routinely hosted phrenological exhibitions. In fact, phrenology was the most popular instructional subject among the Yorkshire Union of Mechanics' Institutes (to which the Keighley Mechanics' Institute belonged), with thirty-two phrenology lectures delivered in 1842 alone (Cooter, *Cultural Meaning* 149).

The reading material to which the Brontës had access also routinely addressed the science. The two periodicals the family received, *Blackwood's Edinburgh Magazine* and *Fraser's Town and Country Magazine*, published numerous articles on phrenology which, for the most part, criticized the science.[6] *Blackwood's* attacked phrenology through satire, and *Fraser's* through a more serious consideration of its materialist implications.[7] Nonetheless, both periodicals significantly contributed to phrenology's rapid cultural diffusion. A particularly dramatic phrenological tale that addressed the science in relation to early education appeared in *Fraser's* popular series "Remembrances of a Monthly Nurse" by Harriet Downing.[8] The nurse narrator, initially skeptical of the phrenological premise that identity has its origin in the material organization of the brain, becomes convinced when the adult idiot son of her employer has the power of his intellectual faculties, always well-formed but somehow impeded, completely restored after sustaining a severe head injury. The unconventional situation leads the nurse to counsel the family to educate the man as if he were a young child, so as to carefully unfold his "yet undeveloped faculties" (508). In terms of books to which we know the Brontës had access, the Keighley Mechanics' Institute library contained Robert Macnish's *Philosophy of Sleep* (1830), which informed Patrick Brontë's annotations on states of unconsciousness in the family's copy of Thomas Graham's *Modern Domestic Medicine* (1826) (Shuttleworth, *Charlotte Brontë* 28).[9] In that volume, Macnish explains that his theories on sleep are based in the "phrenological system," which he avers is "the only one capable of affording a rational and easy explanation of the phenomena of mind" (vii).[10]

But perhaps the most interesting direct connection between the Brontës and phrenology is Charlotte's enthusiastic reaction to the science after having her head read by a professional phrenologist.[11] While spending time in London in 1851, she and her publisher, George Smith, shielded their identities by adopting the pseudonyms Mr. and Miss Fraser and went to the phrenological salon of J.P. Browne. Smith later received Browne's written assessments, prompting Brontë to write asking for her "character" after she returned to Haworth. In the account, Browne makes note of her "large and well formed" intellectual region, her "fine organ of language," and sentiments that are "imbued with that enthusiastic glow which is characteristic of poetical feeling" (qtd. in Smith 787). Smith was not completely satisfied with his reading, but Brontë found it to be wonderfully accurate: "I wanted a portrait, and have now got one very much to my mind . . . it is a sort of miracle—*like—like—like* as the very life itself. Destroy Mr Ford's lithograph. Transfer to fair type Dr Browne's sketch, and frame and glaze it instead. I am glad I have got it. I wanted it" (qtd. in Wise and Symington, 258). Nicholas Dames claims that Brontë's interpretation of this experience "offers us glimpses into a structure of experience in which the immediately visible, the objectified exterior, may be prized over any private interiority" ("Clinical Novel" 367). According to Dames, the widespread tendency for critics to appreciate Brontë's texts for representing interiority and psychological complexity fails to account for the "clinical" discourses of phrenology and physiognomy, sciences that represent a "model of visuality" that values surface over depth (*Amnesiac Selves* 86). Dames aligns phrenology with visual appraisal and positions this interpretive paradigm in opposition to depth psychology, a nonvisual form of assessing character. He is certainly not wrong: Charlotte Brontë's fiction is rife with characters attending to the contours of foreheads, the size of organs, and the general shape of skulls—and her novels consistently endorse the accuracy of this form of character analysis. That said, she does not endorse phrenological interpretations in place of understandings of interiority but, rather, in opposition to *another* form of visual assessment, one that has far higher stakes for women than for men: determining value based on physical appearance. This is even evident in her reaction to the phrenological reading that she asks Smith to "frame and glaze" in place of a lithographic representation: for her, a visual portrait fails where the verbal "sketch" achieves true likeness. This desire to value abstract qualities of mind over physical appearance might also operate as a form of wish-fulfillment on Brontë's part, given how sensitive she was about how she looked.[12] As Smith observes shortly before his report of their phrenological experience in the *Cornhill Magazine*:

> There was but little feminine charm about her; and of this fact she herself was uneasily and perpetually conscious. It may seem strange that the possession of genius did not lift

her above the weakness of an excessive anxiety about her personal appearance. But I believe that she would have given all her genius and her fame to have been beautiful. Perhaps few women ever existed more anxious to be pretty than she, or more angrily conscious of the circumstance that she was *not* pretty. (784–85)

If we credit Smith's account (which Brontë biographer Juliet Barker characterizes as "perceptive") (660), the fact that Brontë was "angrily conscious" of her lack of beauty indicates her dissatisfaction with two conflicting modes of interpretation related to visual assessment: how she wishes others would see her, and how they actually do. That this incongruence provokes anger betrays a sense of injustice—a belief that it is unfair that she is not as beautiful as she feels she should have been. Given that having her portrait professionally made at Smith's insistence occasioned her bursting into tears twice, it is perhaps unsurprising that she would prefer a phrenological sketch to a visual one.[13]

This shift in valuation away from the fetishization of physical beauty and toward an appreciation of mental qualities is a thematic concern in much of Brontë's fiction, and she consistently uses an appeal to innate character to validate romantic relationships based on psychological compatibility rather than physical beauty. This is particularly evident in Jane Eyre's evolving sense of her self-worth in comparison to Blanche Ingram. At first, Jane is convinced that Rochester will prefer Blanche's "lovely face" to her own plain features, going so far as to adopt the "wholesome discipline" of drawing an honest chalk portrait of herself and an ivory miniature of Blanche in order to steel herself against the inevitability of Rochester's choice (137). Eventually, however, she comes to claim her superiority as a partner by elevating the value of mutual understanding over Blanche's superficial attributes. After Rochester's misleading insinuation that he will soon marry Blanche, she indignantly responds:

> Do you think, because I am poor, obscure, plain, and little, I am soulless and heartless? You think wrong! [. . .] And if God had gifted me with some beauty and much wealth, I should have made it as hard for you to leave me, as it is now for me to leave you. I am not talking to you now through the medium of custom, conventionalities, nor even of mortal flesh—it is my spirit that addresses your spirit . . . equal,—as we are!

> [. . .] and yet not so; for you are a married man—or as good as a married man, and wed to one inferior to you—to one with whom you have no sympathy . . . I would scorn such a union: therefore I am better than you. (216)

Shortly before this passage, Jane compliments Rochester for possessing an "original, a vigorous, an expanded mind," which she contrasts with the "inferior minds" she has encountered at Lowood and Gateshead (215). In asserting her equality with him, she affirms her own intellectual power that she then claims entitles her to his regard whether he bestows it on her or not. While Jane is Rochester's inferior in terms of wealth and status, and Blanche's inferior in beauty, a valuation of character inverts conventional hierarchies and places her on top. Jane puts physical appearance (being "little" and "plain") in the same category as social connection and economic standing, despite the fact that having "some beauty" is at least as dependent on biological inheritance as a large organ of adhesiveness. The difference is that feminine beauty is a form of nature that has been unduly privileged over qualities of mind in much the same way as wealth and rank. Beauty is an economy like any other, and in order to be properly valued, Jane asks Rochester to invest in a new market: the currency of character, which pays its dividends in psychological compatibility rather than immediate physical attraction. Jane's campaign (and Brontë's larger project regarding beauty) salvages the apparent sexual sacrifice as well by converting companionability into a different kind of catalyst for desire. Jane has at this point already internalized this lesson, telling Rochester that "A loving eye is all the charm needed: to such you are handsome enough; or rather, your sternness has a power beyond beauty" (209). Jane's claim that "sternness" can be sexy is not a case of beauty being in the eye of the beholder, but rather, that beauty should be beside the point because specific aspects of personality can substitute for physical attractiveness and provoke equal amounts of pleasure. Whether or not this new source of sexual desire has purchase, however, depends entirely on the degree to which these qualities appeal to reciprocal aspects in the character of the lover.

This conversation about the importance of esteeming compatibility over status and recognizing the substitutive value of innate psychology for beauty sets the stage for Rochester's revelatory proposal, "My bride is here . . . because my equal is here, and my likeness" (217). Given her poverty and position, when Rochester says Jane is his "equal," he means they have equivalent powers of mind, and possibly, that they are equally unattractive. While the latter insinuation might seem an unlikely one in a marriage proposal, it is plausible given Rochester's awareness of his own unattractive appearance, which he, too, measures against Blanche. As he explains, "I wish, Jane, I were a trifle better adapted to match with her externally . . . can't you give me a charm . . . to make me a handsome man?" (209). Brontë here foregrounds Rochester's awareness of natural inequalities: Blanche is his superior in beauty if inferior in mind. Rather than engage in the customary and socially expected exchange of male wealth for feminine beauty, Rochester rejects a "purchase" model of matrimony in favor of a union that is not about acquiring something one does not already have, and is based, instead, on enlarging and enriching one's experience of self—obtaining in another one's own "likeness."

Jane Eyre's valorization of natural compatibility over physical attraction mirrors phrenological discourse on selecting a mate. Phrenological texts on marriage routinely stress the importance of finding a partner with a character uniquely suited to one's own physiological traits. As Combe explains, a "natural law in regard to marriage, is, that the mental qualities and the physical constitutions of the parties should *be adapted to each other.* If their dispositions, tastes, talents, and general habits harmonize, the reward is domestic felicity . . ." (*Moral Philosophy* 28). One of the things that obscures a person's ability to properly assess character in the selection process, however, is sexual desire for an attractive partner. Combe notes that people "actuated chiefly by passion, often make unfortunate selections of partners, and entail lasting unhappiness on each other" (26). A similar warning appears in J.C. Lyons's 1846 *The Science of Phrenology as applicable to Education, Friendship, Love, Courtship, and Matrimony* (produced by Aylott and Jones, publishers of the Brontës' poetry). Lyons laments that rather than studying cerebral development, young women are routinely fooled by physical appearance and the artifices of romance: "The generality of young ladies are led away by a handsome face, good figure, warm passion, flattering tongue, and, lastly, a polite address; whilst they invariably neglect the most important consideration—namely, a well-formed head" (16). In the phrenological marriage model, superficial qualities like beauty and charm are unmasked as overvalued and potentially deceptive. The body remains important only insofar as it is an index of the health and character of a potential partner, not as something of value in and of itself.

Jane and Rochester certainly seem to develop their romantic relationship along phrenological lines as they take turns appreciating the others' "well-formed head[s]." In one of their first exchanges, immediately after Jane tells him he is not handsome (and right before he informs her she is not pretty), Rochester lifts his hair to display his "solid enough mass of intellectual organs," and then goes on to point to "the prominences which are said to indicate [conscience], and which, fortunately for him, were sufficiently conspicuous; giving, indeed, a marked breadth to the upper part of his head . . ." (112–13). Rochester offers his head to Jane to challenge her evaluation of his attractiveness on the basis of appearance, and it works: after admitting that "most people would have thought him an ugly man," Jane nevertheless finds that his "reliance on the power of other qualities, intrinsic or adventitious, to atone for the lack of mere personal attractiveness" convinces others to "put faith in [his] confidence" (113). In this scene it seems that Rochester has accomplished what Brontë could not: a pride in his mental abilities so firm that it convinces him and others of his own attractiveness. Jane does note one area for concern in his head, a declivity "where the suave sign of benevolence should have risen," but she is never distressed by this sign (112). The overall effect of Rochester's cranial display on Jane is a recognition and appreciation of his "intrinsic or adventitious" qualities that transform him into a desirable man despite his physical unattractiveness.

Rochester also avails himself of the opportunity to conduct a phrenological reading of Jane, and in doing so confirms that she, too, possesses impressive mental qualities. While disguised as a fortune teller, Rochester performs a prolonged visual assessment of Jane's head, culminating in him ventriloquizing the testimony of her frontal lobe: "that brow professes to say,—'I can live alone, if self-respect and circumstances require me so to do. I need not sell my soul to buy bliss. I have an inward treasure, born with me, which can keep me alive if all extraneous delights should be withheld; or offered only at a price I cannot afford to give'" (171). On one level, the reading foreshadows Jane's decision to reject Rochester's invitation to become his mistress, choosing not to "sell [her] soul to buy bliss." On yet another, however, it reveals how Rochester estimates Jane's worth based on her mental attributes. The organs along the brow are the "perceptive" faculties (also referred to as the "intellectual organs"), and taken together they guide one's experience and understanding of the external world. Thus, what Jane's brow reveals is that the mental lens through which she encounters the world perpetually enriches her experience of it. By saying that her forehead boasts of an "inward treasure, born with [her]," Rochester underscores not only that Jane's mental qualities are considerable but also that these gifts come from nature. Jane's inheritance is physiological and psychological rather than social and monetary, and Rochester's ability to recognize and properly value inherent mental capacities over every other way of estimating a woman's worth is a key way in which Brontë attempts to make their relationship plausible. Being raised out of her station and circumstances is in some measure an external recognition of Jane's internal value—or, as Rochester explains to her, "I will myself put the diamond chain round your neck, and the circlet on your forehead . . . *for nature, at least*, has stamped her patent of nobility on this brow . . ." (220, my emphasis). While in every other way Jane may be poor and lowly, Brontë can cast her as superior by applying nature's rubric rather than society's. In offering to literally crown her, Rochester externalizes his recognition of Jane's internal value, closing the gap between her naturally noble forehead and the mark of status it has hitherto been denied.

There's something surprising in Rochester's support of the idea of a hierarchal structure based in nature rather than the class into which one is born. He is, after all, the beneficiary of the existing economic system and a representative of the elite. While phrenology was the authorizing science of the upwardly mobile working and middle classes, it was actively opposed by many born into socioeconomic privilege.[14] This stands to reason, since the latter had nothing to gain and everything to lose from a science that implicitly supported a reordering of society based on individuals' natural talents and aptitudes rather than lineage. Brontë, however, makes Rochester's support of meritocratic naturalism plausible by making it follow from the disastrous experience of his first marriage. Rochester describes Bertha's head in explicitly phrenological terms, and in a way that mirrors the science's use in racist and proto-eugenic discourses. He tells Jane, "What a pigmy intellect she had—and what giant propensities! How fearful were the curses those propensities entailed on me!" (261). The "propensities" refer specifically to those organs at the back of the skull that relate to self-preservation, survival, and procreation. Combe and other phrenologists repeatedly claimed that European skulls had larger moral and intellectual regions, while the animal propensities were predominant in the "savage" races.[15] As Combe writes in the *Constitution of Man*, "Every phrenologist knows, that the brains of the New Hollanders, Caribs, and other savage tribes, are distinguished by great deficiencies in the moral and intellectual organs" (49). The question of Bertha Mason's racial identity has garnered a great deal of critical commentary, but regardless of whether she is of mixed race, the skull Rochester describes is entirely consistent with non-European racial types in phrenological discourse.[16] By invoking the "curses" of Bertha's propensities, Rochester is appealing to racially inflected biological determinism to underscore the inevitability of future discord with someone possessing a brain so radically different from his own. Or, as he puts it, "her cast of mind [is] common, low, narrow, and singularly incapable of being led to anything higher, expanded to anything larger . . ." (261). The point he is making is not merely that she resisted improvement but that she lacks the mental capacity to improve at all.[17]

Rochester's attribution of Bertha's madness to the "germs of insanity" she inherited from her institutionalized mother also echoes phrenological discourse on hereditary (261). Decades before Darwin's work on inherited traits or Galton's formulation of eugenics, phrenological works emphasized the importance of biological transmission in contributing to the physical and mental development of offspring. As Spurzheim explains, "He who can convince the world of the importance of the laws of hereditary descent, and induce mankind to conduct themselves accordingly, will do more good to them, and contribute more to their improvement, than all institutions, and all systems of education" (*View of the Elementary Principles* 51). Possibly seeing himself in this role, Combe stresses the importance of selecting a spouse with an eye to the progeny one hopes to produce, explaining that "the union of certain temperaments and combinations of mental organs in the parents, is highly conducive to health, talent, and morality in the offspring, and *vice versa*." Failure to consider the organic suitability of their partner for procreative purposes results in "The punishment [of] debility and pain transmitted to the children." He further explains that those who do not consider hereditary influence before marriage are either ignorant of biological transmission or are led by the influence of their "sensual appetite" (*Constitution* 109). This is certainly the case for Rochester, who was in large part induced to marry Bertha for her attractiveness. As he explains, "Miss Mason was the boast of Spanish Town for her beauty: and this was no lie . . . I was dazzled, stimulated: my senses were excited . . . I thought I loved her" (260).

Overwhelmed by her beauty and enticed by her wealth, Rochester twice trespasses against natural law by failing to carefully assess the biologically encoded traits in his potential bride and suffers the consequences.

Rochester's trajectory from Bertha to Jane highlights a shift away from valuing wealth and beauty in a partner and toward an appreciation of innate qualities of mind. The fact that a man already at the top of the socioeconomic hierarchy achieves this awareness is extraordinarily convenient for Jane because through marriage she can be instantly elevated to a status commensurate with her personal assessment of her internal worth. Whereas middle-class phrenologists sought incremental change through social reform, *Jane Eyre* uses marriage to immediately redistribute wealth to the mentally superior through the complicity of an upper-class subject. Admittedly, the novel is critical of the existing class system insofar as it points out, repeatedly and emphatically, the inequities of opportunity and the unfair advantages of unearned and undeserved social privilege. The novel's use of innatist psychology, however, ultimately supports replacing a socially constructed hierarchy with a natural one, in which the new elite are nature's nobility: those with physiologically superior minds.

In *Jane Eyre*, the physiological basis of character bolsters a critique of the class system in this reformist, but not subversive, way, with Brontë focusing more on the unbridgeable gulf separating those with superior mental capacities from those with "inferior" brains. From a class perspective, this aspect of phrenological discourse is complicit in supporting the existing social structure by promoting the idea of personal improvement through self-discipline, while simultaneously reassuring the citizenry that the just organization of society is guaranteed by natural law. Understanding one's innate character and natural abilities, as Jane and Rochester do, promised moderate mobility over time, but the "natural" limits of developmental possibility guaranteed that the underlying system would not be overturned.

However, when it comes to domestic influence in the novel, innate psychology emerges as something that can be altered in astonishing ways. Take, for instance, Jane's ability to discipline, correct, and control Rochester's behavior during the period of their engagement. Jane's regimen of training up his morality and curbing his ardor with her "needle of repartee" yields such excellent results that she stores up the technique for later use in marriage, telling herself, "I can keep [him] in reasonable check now . . . and I don't doubt to be able to do it hereafter: if one expedient loses its virtue, another must be devised" (233, 234). Rochester also expresses appreciation for her disciplining method: "Jane: you please me, and you master me . . . I am influenced—conquered; and the influence is sweeter than I can express . . ." (222). In this way, Brontë romanticizes and eroticizes Jane's ability to bring to heel a belligerent, moody, and sexually aggressive man, reframing domestic influence as a form of intellectual foreplay between mental equals.

The crowning achievement of Jane's corrective influence, however, follows from her refusal to become Rochester's mistress. This act, and her subsequent flight from Thornfield, has a profound effect on his moral character, including the portion that is innately deficient. As aforementioned, Rochester has one defect in his moral organs: the "abrupt deficiency where the suave sign of benevolence should have risen" (112). This is no trivial flaw, as the organ regulates the degree to which one considers and attends to the welfare of others. Combe explains that when well-developed, benevolence "loves for the sake of the person beloved" (*Constitution* 61), and Spurzheim similarly notes that it "leads to the fulfilment of the great commandment, *Love thy neighbour as thyself*" (*Outlines* 45).[18] Jane interprets the declivity correctly since she mocks him by ironically asking whether or not he is "a philanthropist"—a role that would require a particularly large development of the faculty (112). And, indeed, Rochester spectacularly proves his lack of benevolence by failing to consider Jane's welfare, first by attempting to marry her under false pretenses and then by suggesting she become his mistress. In both acts, he places his own desires above the potential moral and social repercussions on Jane (and Bertha). The dramatic change in his moral character and religious perspective, however, follows directly from Jane's refusal and flight. Before leaving, she provides him with a good deal of advice—"trust in God," "Believe in heaven," "live sinless," "strive and endure"—placing a sudden emphasis on spiritual concerns in a way she had not formerly in her narrative of their relationship (270). In leaving, however, she catalyzes a transformation that eventually leads to his repentance and acceptance of God's authority. As he later explains, "I did wrong: I would have sullied my innocent flower—breathed guilt on its purity: the Omnipotent snatched it from me" (380). Brontë conflates Jane's departure from Thornfield with God's will, revealing that in taking a principled stand she was unknowingly fulfilling a divine plan that would later return to her the man she loves, re-Christianized. The statement shows not only Rochester's recognition that his earlier attempts to be with Jane were immoral but also that they were immoral because of their potential negative effects on someone else. At least when it comes to Jane, by the novel's conclusion Rochester has managed to develop his capacity for altruism despite his cerebral shortcomings.

By attending to where in the narrative material signs of character have consequences and where they do not, Charlotte Brontë's ideological uses of physiological psychology come into focus. When it comes to class, she invokes the authority of nature to justify Jane's socioeconomic elevation through natural law. To do this, she emphasizes the more deterministic strain of phrenological discourse, thereby reifying a natural hierarchy of mind in opposition to hierarchies based on wealth, station, or beauty. The essentialist perspective highlights how her mind is superior to the minds of Bertha Mason and Blanche Ingram, and her natural compatibility with Rochester reflects the phrenological view that innate character is the most important concern in a marriage—or, as Jane puts it in the concluding chapter, "we are precisely suited in character—perfect concord is the result" (384). However, when it comes to domestic influence, the novel emphasizes the malleability of innate character, which can be effectively altered by carefully calibrating interpersonal interactions and external circumstances. Whether Brontë leans on the fixed or flexible side of phrenology largely depends on the aspect that most benefits and legitimates Jane's progress. When needed, Jane's moral influence tames Rochester, ultimately catalyzing the process through which he finds eternal salvation. This ideology of influence, however, runs counter to the marriage model advanced in phrenological discourse, which stresses the futility of changing the character of one's partner. It is this negative aspect of the science's treatment of marriage that Anne Brontë mobilizes in *The Tenant of Wildfell Hall* to reveal the problematic implications of the cultural belief in a woman's ability to transform a man.

Much like Charlotte's fiction, Anne's novels reveal her familiarity with the principles, terminology, and social implications of phrenology. In the novel *Agnes Grey* (1847), Agnes frequently describes her trials as a governess through phrenological language. For instance, she laments the harm done to the Bloomfield children when their uncle visits because he "encourag[es] all their evil propensities," thereby reversing "the little good it had taken [her] months of labour to achieve" (41). Similarly, she observes that Miss Murray's "love of display had roused her faculties," leading her to only devote time to studying "the more showy accomplishments" such as singing, dancing, and drawing (58). In *Agnes Grey*, phrenology provides Anne with a convenient vocabulary for describing the unique mental attributes of different children, allowing her to underscore the challenges of providing individualized instruction for multiple students simultaneously. The science also highlights Agnes's frustration over her inability to completely control the environment of her pupils, and how such circumstances interfere with the success of a carefully constructed educational plan—issues to which Anne Brontë returns in *The Tenant of Wildfell Hall* as Helen Huntingdon gradually becomes convinced of the necessity of rescuing her son from his father's corrupting influence.

The progressive position the novel takes in response to the shocking subjects it forthrightly addresses has been well recognized and discussed both in our day and its own.[19] While the novel graphically details Helen's abusive marriage to a depraved, alcoholic man, its most radical aspect is the sympathetic portrayal of a wife's adamant resistance to her husband's authority. As May Sinclair famously observed, when Brontë "slammed the door of Mrs. Huntingdon's bedroom she slammed it in the face of society and all existing moralities and conventions" (vi). Her resistance culminates in the illegal act of escaping with her son after she finally accepts the futility of her attempts to influence his behavior in any lasting way. As Priti Joshi argues, the novel "challenges the central tenet of domestic ideology—women's influence on men" by offering "an excruciating exposé of the utter fictionality of this doctrine" (915). Brontë, however, does not just provide an example of an incorrigible husband capable of resisting the influence of a virtuous wife; she also anchors the cause of his intractability in his innate mental dispositions. In doing so, she undermines a model of sexual complementarity based in compensatory spiritual essentialism with a biological essentialism accessed through phrenology. In addition, she uses the science to challenge a father's right to custody on biological grounds, thereby opposing his legal rights with an appeal to natural law.

As opposed to many other aspects of the science, phrenology's treatment of marriage tended not to emphasize a regimen of gradual improvement through mental exercise and self-control. Unlike most arguments for phrenology's practical application, such as prison discipline or educational reform, its usefulness in spousal selection was self-evident and did not require large-scale institutional change. As aforementioned, the science's immediate personal application in this arena also provided professional phrenologists with a rationale for selling manuals and readings. Delineation booklets, in which a client's mental development was recorded for later study, often opened with an encomium on the importance of phrenology in selecting a mate. As one such publication explains, phrenology "informs lovers that if suitably mated their lives would be rendered happy by the union; but if not organized for each other, they ought not to marry, for their lives would be rendered miserable by such a match" (Wells 9). Adopting a more fatalistic tone, another booklet explains that phrenology's "utility in detecting diseases of the brain and body is of great advantage to persons forming matrimonial engagements," presumably by enabling such persons to avoid unwittingly entering into an engagement with someone who has a latent physiological deficiency (Hallchurch and Thorneycroft 3).[20] As such texts repeatedly stressed, for those who failed to heed nature's warning, the marital consequences could be disastrous.

Like so many initially unforeseen implications of the science, the very aspect that proponents championed to increase its popularity potentially threatened its conventional appeal. What logically follows from the premise of indelible organic incompatibility is that couples who failed to consult phrenology must simply suffer, or seek separation or divorce. As Combe contends in his *Moral Philosophy*, marriage is first and foremost a "natural institution" rather than one defined entirely by "ecclesiastical or civil law" (26). A union between naturally unsuited partners is, thus, an "immoral condition" because it violates the "law of God written in our frames" (28, 29). For this reason, physiologically mismatched couples "ought not to be deprived of the possibility of escaping from the pit into which they may have inadvertently fallen; and not only divorce for infidelity to the marriage vow, but dissolution of marriage by voluntary consent, under proper restrictions . . . should be permitted" (33). Henry George Atkinson, phrenologist and friend of Harriet Martineau, similarly argued for the dissolution of marriage on these grounds, explaining that when "Unions of the sexes are formed by chance, the laws of nature are neglected . . . when individuals are unsuited to each other, it is not God who put them together, but their own ignorance and perversion" ("[On a family]," 171). A similar argument appears in a *Times* article covering the radical public lectures of a Mrs. Chappelsmith, who recommends "allowing divorce to be easily obtained by all parties" because people had not hitherto had the benefit of phrenology to assess compatibility ("Socialism" 6).

Viewed in the context of the physiological basis of character, Helen's decision to marry Arthur Huntingdon emerges as an act in defiance of natural law, for which she predictably suffers. Her narrative reflects two key themes of phrenological discourse on marriage already discussed: first, that the examination of innate character should take precedence over all other considerations in selecting a partner, especially that of physical attraction; and second, that an individual cannot greatly alter or affect the character or behavior of a spouse. This second theme follows directly from the argument for phrenology's usefulness in marriage selection, which depicts a union between incompatible partners as a constant and irremediable struggle. By connecting her depiction of incompatibility and ineffectual influence in marriage to discussions of physiological capacities, Brontë deploys scientific naturalism to morally justify Helen's decision to escape with her son.

Throughout the novel, Brontë provides many indications that Huntington and Helen are not physiologically compatible, usually by emphasizing Helen's comparative mental superiority. Gilbert Markham takes note of Helen's impressive intellectual organs, observing that her "forehead was lofty and intellectual," and later noting her "lofty brow, where thought . . . stamped [its] impress" (47, 56). In contrast to this appreciative response, Markham is immediately suspicious of Huntington when he sees his portrait, writing that his "wavy curls, trespassed too much upon the forehead, and seemed to intimate that the owner thereof was prouder of his beauty than his intellect—as perhaps, he had reason to be;—and yet he looked no fool" (71). Markham reads Huntington's appearance as evidence of two possible motives: either the subject is proud of the wrong physical attribute, choosing to conspicuously display his hair rather than his intellectual organs, or he is knowingly concealing his forehead and emphasizing his physical beauty as a way to draw attention away from a lack of intellectual ability. Both possibilities offer evidence of a moral failing, either vanity or cunning. Markham's negative visual assessment of Huntington echoes that of Milicent Hargrave, who questions the wisdom of Helen marrying him based on the evidence of "his look." She comments, "of course he is [handsome], but *I* don't *like* that kind of beauty; and I wonder that you should . . . there's nothing noble or lofty in his appearance." She further explains that when it comes to her taste in men, the "spirit must shine through and predominate" (171). With no explicit reference to the shape of his skull, Milicent's assessment of Arthur's appearance nevertheless makes use of phrenological terminology: when she states the "spirit must shine through," she refers specifically to physical bearing ("appearance"), most likely to the large cluster of moral sentiments that stretch from the upper forehead to the back of the crown (among them "Veneration," "Benevolence," and "Ideality"). By "predominate," she means that these moral qualities should be noticeably larger than the baser tendencies (or "animal propensities") located at the back of the head.[21] In Milicent's visual estimation, Arthur's appearance does not display an abundance of moral tendencies (i.e., creating a protuberance at the crown), although whether or not he is actually deficient in these qualities remains, at this point, a mystery concealed by his abundance of curly hair. Regardless, she feels she has evidence enough to tell Helen, "you are so superior to him in every way . . ." (171). Markham and Milicient's observations mark their similar apprehension of the vast difference between the physical indices of character in the two, emphasizing their innate incompatibility.

Rather provocatively for the time, Helen's desire to marry Huntington stems almost entirely from physical attraction. Understanding passion as a biological phenomenon, phrenology characterized initial attraction between the sexes as a state in which mental organs connected to the instinct to procreate would temporarily override an individual's reason, creating the illusion of deep-seated affection. For the benefit of unmarried women, Combe provides phrenological descriptions of a thief and a rapist he visited in Newcastle Prison. He speculates that women would likely find the men appealing as husbands because both were "well-looking and intelligent in their features" and only later would discover their true characters "after having become their victim" (*Moral Philosophy* 30). Such commentaries make a distinction between two kinds of material signs: the physical attractiveness that appeals to one's sexual instinct and indices of specific characterological traits encoded in the brain. Although both signs are, in the most literal sense, "superficial" in that they deal with the assessment of physical traits, the former signifies only itself (i.e., the state of being physically attractive) whereas the latter is the sign of something more—innate tendencies, predilections, and future behavior. A handsome face and good figure appeal only to the baser animal instincts, whereas the mind has the power to appeal to the moods, tastes, thoughts, and feelings associated with the intellectual and moral faculties. Attention to good looks inspires short-lived passion, but attention to biologically encoded tendencies ensures an enduring sympathy of minds.

Judging from her diary, it is difficult to assign any reason for Helen's decision to marry Huntington other than physical attraction. She defends Arthur to her aunt on the grounds that she "cannot believe there is any harm in those laughing blue eyes," and later writes that she need not record their conversation because it would signify little without the "aids of look, and tone, and gesture" that "made it a delight to look in his face, and hear the music of his voice, if he had been talking positive nonsense . . ." (136, 142). Helen's tendency to focus on Arthur's physical presence rather than his character is further emphasized through the sketches she furtively makes of him. The fact that their tumultuous courtship is mediated through these representations—which Helen attempts to hide, but which Huntington exults in revealing—makes clear not only his vanity but also Helen's desire to repress the true source of her ardor.[22]

Although Helen's physical attraction to Huntington is the primary reason for her decision to marry him, an abetting rationalization is her confidence in the efficacy of domestic influence. The novel makes it quite clear (almost didactically so) that one of Helen's most grievous errors is buying into the ideological marriage model that stressed the redemptive power of self-sacrificing love to influence a man to change. As Mary Poovey has shown, the feminine domestic ideal was product and producer of bourgeois power precisely because it seemed to exist outside the system it helped to maintain. The binary opposition between wage-earning and unpaid domestic labor coded the private sphere as an illusory alternative to male middle-class economic strife, and "the prize that inspired hard work, for a prosperous family was the goal represented as desirable and available to every man" (10). This is an extension of Nancy Armstrong's argument that domestic fiction naturalized the narrative that feminine purity could "nudge sexual desire into conformity with the norms of heterosexual monogamy" (6). Although largely concerned with the domestic sphere, *Wildfell Hall* is the antithesis of domestic fiction as Armstrong defines it because the arc of the narrative, taken as a whole, opposes the doctrine of feminine influence.

Before her marriage, Helen uses this ideology to counter her Aunt's objections to Huntington's character, expressing her enthusiastic desire to improve him with missionary zeal. Repeatedly using the language of salvation, she insists, "I think I might have influence sufficient to save him from some errors, and I should think my life well spent in the effort to preserve so noble a nature from destruction" (146); "I will save him from [his companions]" (147); "I shall consider my life well spent in saving him from the consequences of his early errors, and striving to recall him to the path of virtue" (147). The ideology of influence initially appeals to Helen because it provides her with a quest, positions her as hero, and perhaps most importantly, allows her to disguise her sexual desire as a willingness to undertake a Christian mission for the sake of Arthur's salvation. Such rhetoric at once elevates the wife's importance while paradoxically making her glory conditional upon self-sacrifice. Helen's characterization of such an endeavor as a "life well spent" fittingly encapsulates the relational dynamics that underpin such an ideal: the wife "spends" herself, giving her husband her time, effort, and energy. The asymmetrical transmission that she idealizes as self-sacrifice is an act of self-abnegation—an emptying out of self to increase the value of another. Helen, however, revels in increasing and storing up her own self-worth before marriage so she might later transfer it to her husband: "whatever skill or knowledge I acquire is some day to be turned to his advantage or amusement; whatever new beauties in nature or art I discover, are to be depicted to meet his eye, or stored in my memory to be told him at some future period" (148). Given Helen's repeated, emphatic, and effusive expressions of confidence that she *can* "save" Huntington (undercut throughout by her aunt's skepticism), her self-assurance on this front is likely the "very natural error" Brontë claims, in the preface to the second edition, her novel attempts to prevent young girls from making (40). Helen's choice is based on sexual attraction, but the ideology of feminine influence serves both as a safeguard against error and a way to elevate marriage from a domestic arrangement to a divine mission.

Helen's conviction that she can effect a change in Arthur's character sets the stage for the most dramatic phrenological scene in the novel, in which her husband reveals for the first time precisely how defective his moral region actually is. In a scene that is the inverse of Rochester's proud phrenological display, Arthur uses the sign of his biological deficiency to justify his *carpe diem* lifestyle and irreligious attitude. In the middle of an exchange in which Helen urges her husband to improve his piety, he removes his hat, and counters, "But look here, Helen—what can a man do with such a head as this?" Helen reflects, "The head looked right enough, but when he placed my hand on the top of it, it sunk in a bed of curls, rather alarmingly low, especially in the middle." Helen counters his reasoning by pointing out that while he is clearly not fitted to be a "devotee," he nevertheless has "the capacity of veneration, and faith and hope, and conscience and reason [all specific phrenological organs], and every other requisite to a Christian's character, if [he] choose[s] to employ them," since every faculty "strengthens by exercise" (191). In one of the few treatments of this scene, Marianne Thormählen makes a convincing case for the probability of Anne Brontë alluding to a similar argument made by Mrs. John Pugh in *Phrenology Considered in a Religious Light* (1846).[23] As Thormählen points out, Helen's position echoes that of Pugh, eschewing biological predestination in favor of a phrenological perspective that emphasizes free will in choosing what one does with his or her God-given faculties (however impoverished they may be). Mrs. Pugh's perspective (and Helen's at this point) are of a piece with the dominant view among Christian phrenologists, based in Edinburgh and led by George Combe; however, Arthur's view is very much in line with the materialist phrenologists, based in London and led by Dr. John Elliotson.[24] This latter camp took quite seriously the implications of mind as matter, and concentrated more on the philosophical and social implications of materialism (some leaning toward atheism) rather than concentrating on self-improvement. Arthur's short move from a materialist premise to the idea that Christianity may, in fact, be "a fable, got up by the greasy-faced fellow that is advising me to abstain, in order that he may have all the good victuals to himself" (192) echoes the unorthodox position of the more radical phrenologists, and represents how, as Roger Cooter puts it, phrenology was "widely perceived by [some] as entirely materialist and God-denying" (*Cultural Meaning* 5).

Presented in the first diary entry after they have wed, this exchange marks a significant turn in the text. Helen's reflections on it, in fact, lead to her first serious misgivings about the wisdom of having married such a man. This is also the first time that Helen considers the possibility that there may be limits to wifely influence. Although phrenological texts advertised the tantalizing possibility that one could become the best version of oneself through the sort of faculty exercise Helen urges Arthur to adopt, these same works also routinely emphasized a hard limit to the effectiveness of environmental influence. In fact, these texts often present the failure to condition oneself or others as the primary mystery of identity that phrenology solves. Mrs. Pugh's book opens with an account of her trials as a teacher, having success with some students, but little to none with others. She laments: "How many weeds have I seen grow up, notwithstanding all my attempts to eradicate them! The vain girl is now the vain woman. The perverse and selfish disposition is such still, modified, it may be, by other feelings and surrounding circumstances. In all my pupils, where I have had opportunity of knowing, character remains essentially the same" (5–6). This less flexible determinism, which concedes that personality in most is "essentially" fixed, is precisely what Helen comes to discover in her marriage, thereby giving the lie to the efficacy of domestic influence and revealing feminine self-sacrifice to be an utter waste of energy and resources.

At the level of plot, Brontë's depiction of feminine domestic influence is straightforward: for Helen, it simply does not work. What is of far more importance is the narrative's engagement with two very different models of mental development: one that attributes character entirely to environmental influence, and another that places hard limits on influence's ability to change another. Before her marriage, Helen attributes individual identity almost entirely to the effects of nurture, but after a few months of marriage she slowly comes to recognize nature's determining power in forming and fixing human character. Despite her best efforts, Helen fails to influence Arthur in any way, and in fact, he becomes worse. The entry on their "First Quarrel" illustrates the environmentalist techniques of her carefully calibrated approach in attempting to elicit contrition for callously defending his first mistress to his wife. For a night and the greater part of the next day, she resists engaging with him, noting, "if I began, it would only minister to his self-conceit, increase his arrogance, and quite destroy the lesson I wanted to give him" (196). Interestingly, this comment follows a long and detailed explanation of how, "to [her] private satisfaction," Arthur's environment during his "lesson" is completely controlled: she notes there were no engaging letters, no woman with whom he could flirt, and no possibility of him riding out of town due to the inclement weather. Helen's belief that this plan will work is based on a larger assumption about human psychology: that individual identity is the sum total of experience, and that by controlling those experiences one may control character development. And yet, even under the most optimal conditions for Arthur's reeducation, the day's lesson ends with him calling her something that "sounded very like 'confounded slut'" (197). Although he remains under the sustained influence of Helen, his "angel monitress," Arthur rapidly and spectacularly degenerates (186).

Throughout her diary, Helen becomes increasingly skeptical about the degree to which a husband may be affected by a woman's moral influence. In a conversation between Helen and Milicent in which the latter earnestly expresses her belief that through her efforts her own husband will improve, Helen responds, "He may. . . . Excuse the faintness of my acquiescence . . . but [my hopes] have been so often disappointed, that I am become as cold and doubtful in my expectations as the flattest of octogenarians" (250). Similarly, when little Arthur asks if his father knows better than to tell people to be damned, she muses, "Perhaps he does," and when her son counters, "you ought to tell him, mamma," she wearily replies, "I *have* told him" (309). Helen's half-hearted concessions to the possibility of change through influence—that "perhaps" he knows what he has been emphatically told, that Hattersley "may" improve under Milicent's moral tutelage—contrast markedly with her earlier unbounded confidence. Strangely enough, Arthur has the final word on this struggle between marital nurture and nature when, in a fit of exasperation during his illness, he observes, "of course I ought to be melted to tears of penitence and admiration at the sight of so much generosity and superhuman goodness,—but you see I can't manage it" (361). Putting a fine point on it, Arthur reveals that while perfectly aware of the systems in play, his inability to capitulate is not a matter of choice but of mental incapacity.

Helen's disenchantment with the effects of wifely influence coincides with a shift in her focus from father to son. Despite its emphasis on Arthur's inability to change, the novel's position is not a wholesale rejection of the notion of interpersonal influence and personal development. After all, the preventative conditioning Helen exhibits in her child-rearing practices prove effective by the novel's end. This raises an important question: why does Arthur resist Helen's continual and extraordinary efforts to influence him, while little Arthur does not?

The novel's position on this issue is best understood in terms of a kind of flexible determinism that becomes less flexible through time. It is her son's close resemblance to his father that in Helen's mind morally justifies her decision to leave with her son. In other words, Helen's actions are necessary because biologically little Arthur is just what his name implies—a *little* Arthur. Paradoxically, the stakes of interventionist environmental influence are higher because environmental influence does not account for the origin of his character: little Arthur's mind is not a *tabula rasa*, but a *tabula totus*, already filled with nascent forms of his father's proclivities and predispositions. Helen's awareness of the physical and psychological resemblance between the two is stressed throughout: Helen describes her child as the "tiny epitome of its father" (217), corrects

Walter Hargrave's assertion that the infant resembles her by gloomily saying, "You are mistaken there; it is its father it resembles" (225), and later remarks on how little Arthur's's "bursts of gleeful merriment" alarm her because she "see[s] in them his father's spirit and temperament" which makes her "tremble for the consequences . . ." (282). The similarities between the two heighten the peril of little Arthur's situation by underscoring how he is starting out in life with a biological disadvantage—the natural tendency to develop in the direction of his father.

Helen's increasing concern with the psychological similarities between father and son resembles phrenological interest in the role of hereditary influence on character formation. While making clear that the boy resembles Arthur, the novel also repeatedly stresses that the son's congenital temperament is not as bad as his father's. Likening his innate character to a garden, Helen comments, "Thank Heaven, it is not a barren or a stony soil; if weeds spring fast there, so do better plants. His apprehensions are more quick, his heart more overflowing with affection than ever his father's could have been; and it is no hopeless task to bend him to obedience and win him to love . . ." (313). Helen's assertion that her son's apprehensions and affections are better than his father's "ever" were points to an innate difference, suggesting that the former might be improved in ways the latter never could be because, in Arthur, those same capacities were impoverished from birth. In the novel's conclusion, Gilbert underscores this natural difference, noting that the "mother's image [was] visibly stamped upon his fair, intelligent features," a recognition that echoes his earlier appreciation of Helen possessing a "better furnished skull . . ." (398, 98).[25] Helen's superior physiological constitution, it seems, has a mediating influence on the constitution little Arthur has inherited from his father, which is consonant with the phrenological belief that children inherit traits from both parents.[26] Little Arthur is, on the one hand, hereditarily disadvantaged, but on the other, not a lost cause. It is precisely by positioning him between these two poles that Brontë amplifies Helen's formative significance to nearly epic proportions.

Helen's fervent response to the diagnostic assessment of her son differs sharply from the attitude of Charlotte Brontë's educators, who use physical signs of character to identify and class students. *The Professor*'s William Crimsworth, in particular, relishes his ability to analyze the character of his students by the size and shape of their skulls. He especially lingers over three "specimens" (82) whose cranial developments reveal increasingly egregious psychologies, culminating in that of Juanna Trista, for whom Crimsworth provides a detailed phrenological delineation: "She had precisely the same shape of skull as Pope Alexander the sixth; her organs of benevolence, veneration, conscientiousness, adhesiveness were singularly small, those of self-esteem, firmness, destructiveness, combativeness preposterously large; her head sloped up in the penthouse shape, was contracted about the forehead and prominent behind . . . narrow as was her brow it presented space enough for the legible graving of two words, Mutiny and Hate" (84). The profile of Pope Alexander the sixth was a frequently used phrenological illustration of inveterate vice (Figure 2.1).

Figure 2.1. Illustration of the head of Pope Alexander VI from J.G. Spurzheim's *Phrenology in Connexion with the Study of Physiognomy*, London, Treuttel, Wurtz, and Richter, 1826.

As Spurzheim writes of the pontif, "such a brain is no more adequate to the manifestation of Christian virtues, than the brain of an idiot from birth to the exhibition of the intellect of a Leibnitz or a Bacon. [. . .] The sphere of its activity does not extend beyond those enjoyments which minister to the animal portion of human nature" (*Phrenology* 71). Spurzheim's analysis, quoted at length in Combe's *Constitution of Man*, resembles Crimsworth's further description of Juanna, whose "animal" nature manifests itself in spitting, making noises like a horse, and relishing her future in the West Indies where she will have access to slaves she can physically abuse at will (84–85).

While Juanna is anomalous in her extreme depravity, Crimsworth makes clear that the majority of the students in both M. Pelet's and Mlle Reuter's school are dim-witted, vulgar, and unmotivated—characteristics that he ascribes almost entirely to their nationality. He identifies Pelet's school as "merely an epitome of the Belgian Nation," and writes of the students, "Their intellectual faculties were generally weak, their animal propensities strong; thus there was at once an impotence and a kind of inert force in their natures. . . . Such being the case it would have been truly absurd to exact from them much in the way of mental exertion . . ." (57, 56). Crimsworth's success as a pedagogue has less to do with his teaching than with taxonomy: by recognizing the futility of meaningfully educating his students, he can see his way to embracing a system of mass psychological supervision that at least curtails the tendency toward group defiance that had led to the dismissal of prior masters (56).

Crimsworth's ability to estimate the average intelligence among many individual heads in a classroom replicates, at a smaller scale, the phrenological project to develop a racial typology based on cerebral differences. *The Phrenological Journal* routinely published articles devoted to analyzing the heads of specific racial or ethnic groups.[27] By 1848, the subject was popular enough to spawn a monthly periodical called *The Ethnological Journal; A Magazine of Ethnography, Phrenology and Archaeology, Considered as Elements of The Science of Races*. Although Combe mentioned race in many of his works, he included a substantial section on the natural differences in national character in his *System of Phrenology*, which he based on his analysis of the collection of international crania held by the Phrenological Society of Edinburgh. Writing against the "fashionable doctrine" that environmental influences create differences in national character, Combe insists that a "common form" and "common proportion" characterize the skulls of each nation or tribe (420, 424).

Typical of such analyses, the study confirms the comparative superiority of European brains in general, and British brains in particular. While the racist implications of such studies is clear, the Scottish phrenologist nevertheless opposed slavery and imperialism, albeit for practical reasons stemming from the internal logic of his science. The difference between European and other races was considered too great a gulf to be bridged by hybridization or enculturation, rendering colonization illogical, and engaging in slavery was considered an unwholesome abuse of the moral faculties. As one historian puts it, "This was not so much the racialism of hatred and contempt as it was the racialism of self-superiority and aloofness" (De Giustino 71).

This description well characterizes Crimsworth's perspective, as he tempers prejudice with humanitarian concern while also making clear the marked difference between himself and the traits common to other groups. For instance, when he observes Pelet's unkind behavior toward two Flemish ushers, he muses that while "intellectual inferiority is marked [on them] in lines none can mistake; still they were men . . . and I could not see why their being aboriginals of the fat, dull soil should serve as a pretext for treating them with perpetual severity . . ." (58). While he disapproves of outright bigotry, he nevertheless matter-of-factly acknowledges the visible signs of the men's inherent inferiority, yoking their individually displayed traits to an ethnic identity defined, not by a common culture but by hereditary descent within a shared environment. As Nancy Stepan has argued, the point at which phrenological science tipped from being politically reformist to forming the conservative foundations of a racial biology is precisely in its move from individual to group traits (23–24). These two paradoxical aspects of the discourse surface in the novel, breaking precisely along the individual/group divide: while Brontë casts Crimsworth as uniquely intelligent and perceptive owing to his innate qualities of mind, the indigenous people with whom he must interact form a single "type" defined by shared biological characteristics.[28] Much like Jane Eyre, Crimsworth is a member of nature's nobility, gifted with "bumps of ideality, comparison, self-esteem, conscientiousness"—but unjustly barred from his proper sphere by family and circumstance (23).

Sally Shuttleworth and Alan Rauch have demonstrated the ways in which Crimsworth uses his phrenological knowledge to gain power and mastery over others despite his initially precarious economic and social position. As Shuttleworth explains, the extensive descriptions of his students' cranial developments function to "shift the boundaries of self-definition from the problematic arenas of class and gender to the more simplistic registers of race and sex" (*Charlotte Brontë* 135). Indeed, the Belgian backdrop of dimwitted boys, cunning girls, and dull ushers allows Crimsworth to foreground his own mental superiority by pointing out the inferiority of others. While these groups certainly serve a purpose for Crimsworth, they get very little in return beyond a regimen of moral management and discipline. As a pedagogue, Crimsworth is not particularly invested in the intellectual improvement of the majority of his students, not because he lacks skill as a teacher but because he has recognized the futility of trying to better minds incapable of significant advancement. The exception that proves the rule is Frances Henri, whose education he encourages and assists because he recognizes her innate abilities: "The shape of her head . . . was different, the superior part more developed, the base considerably less—I felt assured at first sight that she was not a Belgian . . . the type of another race . . ." (102). Frances presents the inverse cranial shape of Juanna, with large intellectual faculties and small animal propensities, and she is not only not Belgian—she is half English. Crimsworth's use of the word "type" to retrospectively describe his future wife's innate psychology reflects the word's secondary meaning of a singular and exemplary instance of an ideal. And indeed, she later confirms his early assessment by becoming his "treasure" and "best object of sympathy on earth" (141). Within the world of the narrative, both Crimsworth and Frances emerge as remarkable individuals because they are not part of a group defined by shared traits: their particularity depends upon their proximity to the sameness of an inferior average.

Although Charlotte Brontë sets *The Professor* in a school with a teacher as the protagonist, she most notably uses those aspects of phrenological discourse related to race and class. In contrast, Anne Brontë's use of phrenology in an educational context more closely aligns with its actual use in literature on child development and application in schools. As Stephen Tomlinson has shown, phrenology played a central role in mid-century educational reform at the national level in both Britain and America and was also instrumental in pushing through the 1870 Education Act (177–205, 316–22). To enact and demonstrate the efficacy of their approach, phrenologists founded model schools throughout England and Scotland, such as the Williams Secular School and the Edinburgh Model Infant School (De Giustino 168, 175). The curriculum in phrenological schools was more "practical" in character, emphasizing science and mathematics over classical education, and object lessons over rote learning

(166–74). The highest priority for phrenological educators, however, was improving the character of students, which they believed would lead to a decrease in crime and general improvement in the intellectual and moral character of the British population (168). To effect this change, phrenological pedagogy supported an individualized diagnostic and prescriptive method, in which each child's innate characteristics were assessed to tailor lessons to their unique aptitudes.

James Simpson, who used phrenological principles to justify a campaign for a national system of secular public schools in the 1830s, outlines the strategies of this character-centered approach to education in his extremely influential *Philosophy of Education*. In order to effect an improvement in the psychology of children, Simpson lays particular stress on moral training, "education's paramount object" (45). Cultivating the sentiments and suppressing the animal propensities was most important for children under the age of six, after which time making a substantial change in a child's disposition was "nearly hopeless" (110). Moral education takes the form of encouraging and applauding demonstrations of ethical behavior and making an appeal to the higher faculties when a child acts immorally (97–98). Simpson's method requires the instructor to constantly monitor the child's experience of the world since the infant's mind would not be developed enough intellectually to learn by book or precept (101–3). Thus, the child's immediate environment, interactions with others, and every activity assumes an incredible importance. The extraordinary degree of supervision required to properly direct the moral development of a child is brought home by Simpson's inclusion of a model "Record of Duties," which he recommends mothers and teachers use to make a daily assessment of their charges (Figure 2.2).[29]

The table includes a lists of duties that correspond to specific phrenological organs, and a register in which to record the degree to which each duty was observed ("O" for obeyed, "N" for neglected, etc.). The record, which Simpson claimed was in regular use by families throughout England and Scotland, was but one tool in Simpson's so-called "MONITORIAL SYSTEM," a process that demanded constant, vigilant, and scrupulous attention to a child's environment and behavior.

The phrenological program of individualized assessment and customized instruction aimed at raising the moral and intellectual character of every student differs enormously from Crimsworth's method. While skilled at visually analyzing his students' character, his pedagogical practice addresses each group as a single category defined by shared characteris-

(237)

No. III.

SPECIMEN OF THE DAILY RECORD OF DUTIES, ORGANIC, MORAL, RELIGIOUS, AND INTELLECTUAL, AS KEPT FOR ONE WEEK.

			Sun.	Mon.	Tues.	wed.	Thur.	Frid.	Sat.
		Organic Duties.							
		Moderate and Wholesome Food,	O	T	S T				
		Air and Exercise,	O	N	—				
		Cleanliness,	O	N	T	—			
		Early Hours, but sufficient Sleep,	O	N	—				
		Moral and Religious Duties.							
Regulation of the Propensities.	7.	Gentleness, Forbearance, no Contention,	O	W	T	S T			
		Courage, no Cowardice,	O	W	V W	N	—		
		Activity, no Listlessness or Idleness,	O	N	T	W	—		
	8.	Good Temper, no Passion or Cruelty,	O	T	S T				
	9.	Openness, no Cunning or Deceit,	O	N	T	S T			
	10.	Frugality no Greediness, or Miserliness,	O	N	T	S T			
		Humility, no Pride, no Meanness,	O	T	S T				
	12.	No Insolence, Derision, or Provocation,	O	T	S T	—			
		No Self-Prefer., no Jealousy, no Envy,	O	T	S T	—			
	13.	Regard to good Opinion, no Shamelessness,	O	T	S T	—			
		No Courting of Praise, no Vanity,	O	T	S T	—			
	14.	Caution, Circumspection, no Rashness,	O	N	T	—			
Exercise of Moral Sentiments.	15.	Spontaneous Kindness, no Coldheartedness,	O	N	W	V W	—		
	16.	Truth, Justice, Charitable Judgment, Candour, Gratitude,	O	T	S T	W	S T		
		Conscientious Duty, seen or not seen,	O	N	W	T			
		Love and Obedience to God,	O	N	T	S T			
	17.	Religious Duties,	O	N					
		Obedience and Deference to Parents,	O	T	S T				
		Respectfulness to Super., Equals, Inf.,	O	T	S T				
	18.	Cheerfulness, Content,	O	N	—				
	19.	Fortitude, Resistance of Temptation, no obstinacy,	O	N	T	S T	W		
	20.	No Exaggerat. or Marvellous Embellish.	O	T	—				
	21.	Refinement, no Vulgarity,	O	T	—				
		Intellectual Duties.							
		Accurate Obser. of Objects and Events,	O	N	W	—			
		Attentive Study and Improvement,	O	N	W	—			
		Order and Punctuality,	O	N	—				
		Exercise of Reflection and Good Sense,	O	N	W	—			

EXPLANATION.—The figures on the left denote the Faculties concerned in the duties (see Table, p. 110). The pupil enters in pencil, to be inked, if approved by the teacher, the fulfilment, &c. of each duty, thus,—by the letter O, if obeyed,—N, if neglected,—T, if transgressed,—ST, if seriously transgressed. The mother, or teacher, alone, enters W for well done,—VW, very well, when respectively merited. The hyphen or score means no entry called for. Each book, in quarto size, exactly like the well known annual house-book from which it was copied, lasts the pupil a year.

Figure 2.2. Record for a child's daily duties from James Simpson's *Philosophy of Education with Its Practical Application to a System and Plan of Popular Education as a National Object*, A. Edinburgh and C. Black, 1834.

tics rather than a collection of unique individuals. Because the Belgian students are, for him, essentially all the same, he can treat the many as one and absolve himself from undue effort. For his male students, he "br[ings] down [his] lesson to the lowest level of [his] dullest pupil's capacity," and regarding the female students he observes that "where the temperament is serene . . . the intellect [is] sluggish, and unconquerable dullness opposes every effort to instruct" (56, 100). Assessing the innate character of his pupils is not a starting point for improvement but rather a means by which he can obviate useless labor and instead direct his energy toward Frances, the one exceptional individual who deserves his instruction because she alone has the capacity to improve.

In contrast, Helen's investment in diagnosing, monitoring, and influencing her son's moral development reflects the child-centered approach of phrenologically based education. While the science was widely used in schools, it was also recommended to mothers since they served as the primary caregivers of children during the time in which the moral organs were most susceptible to improvement. Opposing the notion that women's primary role was to be a domestic ornament to entertain and amuse their husbands, Combe argued that women's primary function was in monitoring and improving the physical and mental condition of future citizens. To achieve this, he recommended that women's education be expanded to include "every species of knowledge," including physiology, mental philosophy, and human anatomy (*Lectures* 50, 52–55). Although Combe, like most popularizers of phrenology, was not motivated by any progressive interest in women's rights, his concerns accorded women more authority and more educational access on biological grounds. While such perspectives still locate the woman's role in the domestic sphere, they nevertheless advocate for an increase in women's authority in the home and greater access to knowledge.

Women who wrote popular advice literature on child rearing frequently used phrenological principles to foreground the mother's importance in character development. The prolific and prominent educational writer Louisa Barwell championed the utility of phrenology in her *Nursery Government* (1836), Mrs. Thomas Spurr applies phrenological principles to improving children's brains in her *Course of Lectures on the Physical, Intellectual, and Religious Education of Infant Children* (1836), and Elise Von Lersner advocates the use of phrenology in her *Children's Gifts and Mothers' Duties* (1865). As the latter author observes, a mother's first duty is to carry out "the mental and physical education from the first period of [a child's] young life" by "keeping from the child everything to which it should not be accustomed, and to repeat and keep before it, that which should become its second nature" (30). In one chapter, Lersner imagines the initial horror a mother would likely experience if she found her infant to possess a skull similar to that of Pope Alexander VI (18–24). She explains that such a case, though dire, is not without hope if the mother redoubles her watchfulness, engagement, and moral protection: "you must encompass him with a shield that shall preserve his innocence" (22).

Tenant also endorses this perspective, which ultimately works to validate Helen's decision to leave her husband for the sake of her son. Helen's decision to shift her focus from the father to the son coincides with her recognition that while her husband is incorrigible, her son is still capable of being beneficially influenced. This, however, is repeatedly overturned by the father's counteractive interference. She writes that when Arthur is away, she has a "chance of recovering the ground [she] had lost," but as soon as he returns, he does "his utmost to subvert [her] labours and transform [her] innocent, affectionate, tractable darling into a selfish, disobedient, and mischievous boy; thereby preparing the soil for those vices he has so successfully cultivated in his own perverted nature" (282). The description casts the couple as outright opponents over disputed ground, and the propagative metaphor underscores the importance of early cultivation. Helen's focus at this point is not on the vices themselves but on the psychological conditions that give rise to future immoral behavior. Because her son's nature is not "a stony soil" like his father's, he still has a chance of being influenced in ways his father could not (313). However, the son's biological inheritance from his father raises the stakes of the cultivation contest between husband and wife. Helen writes, "not only have I the father's spirit in the son to contend against, the germs of his evil tendencies to search out and eradicate, and his corrupting intercourse and example in after life to counteract, but already *he* counteracts my arduous labour for the child's advantage, destroys my influence over his tender mind, and robs me of his very love . . ." (281). Helen's position becomes untenable because two forms of influence, the biological and the environmental, have combined in such a way that she cannot successfully discharge her moral and maternal duty unless she removes her son to a more controlled environment. The hereditary component, figured as the "germs of [Arthur's] evil tendencies," figures importantly in this decision because it tips the scales in the father's favor in the battle of parental influence.

Superadded to this congenital tendency toward general vice is Helen's more specific fear concerning her son's possible future addiction to alcohol. Much has been said about Brontë's portrayal of Arthur Huntingdon's drinking, particularly as a symptom of his vexed relationship with class and gender roles.[30] In addition to its connection to these social systems, however, contemporary medical and scientific texts frequently depict alcohol addiction as an innate tendency, often with recourse to phrenology. As Marianne Thormählen has observed, Arthur bears a strong resemblance to the "sanguineous drunkard" described in physician Robert Macnish's popular treatise *The Anatomy of Drunkenness* (1827): "Persons of this stamp have usually a ruddy complexion, thick neck, small head, and strong muscular fibre. [. . .] In such people, the animal propensities prevail over the moral and intellectual ones. They are prone to combativeness and sensuality, and are either very good natured or extremely quarrelsome" (53). Like this species of drunkard, Arthur also has a red face, and Helen describes him early on as possessing a "sanguine temperament" (146). As a phrenologist, however, Macnish's comment that in drunkards "the animal propensities prevail over the moral and intellectual ones" indicates a biological component to drunkenness located in the brain. This, too, appears to match Arthur's physical description insofar as his suspiciously low curls (cloaking his intellectual faculties) and the coronal depression over his moral organs indicate that in his skull the animal propensities also predominate—a trait further corroborated by Helen's observation that he is "given up to animal enjoyments" (220). An oft-cited study on intemperance by the phrenologist and physician Charles Caldwell similarly indicates that "drunkenness consists in a morbid condition of the brain," located in the organ of "alimentativeness," which corresponds to a person's capacity to enjoy food and drink (323, 330–33). He concludes that if there is a very strong congenital predisposition to drink, an individual will become "the habitual drunkard, whose fiery appetite drags him to the bottle . . ." (333).[31] Such accounts, owing to the presumed connection between the physiological structure of the brain and a pathological behavior, portray addiction as something beyond the control of the inveterate drunkard. Spurzheim, in fact, identifies habitual drunkenness as one of those cases "in which it is extremely difficult to decide whether there is or is not will" (*View* 303). This depiction of the drunkard who remains passive in the wake of an inborn desire resonates with Arthur's self-description after he tosses back one of several tumblers of wine and declares, "I have an infernal fire in my veins, that all the waters of the ocean cannot quench!" (227) Helen writes that she was about to ask "'What kindled it?'" before being interrupted by the butler. This unanswered question about causality rests at the very center of Helen's predicament, because if something innate creates the need that drives Arthur's actions, then it becomes less likely (if not impossible) that he can change his own behavior—let alone that she can influence him to change. Much like the scene that highlights Arthur's deficient moral organs, the narrative again focuses on a physiological trait as a way of accounting for internal motivation that may be beyond one's ability to regulate.

Helen clearly does come to believe Arthur's intemperance is congenital since she fears her son may have inherited the same predisposition. Because alcohol addiction was a clearly identifiable pathology expressed through habitual behavior it was, along with insanity, a trait that phrenologists used to provide evidence for the hereditary transmission of psychological traits. Combe uses intemperance as his prime example of the "Hereditary Transmission of Qualities" in *The Constitution of Man*, relating the story of a Russian family in which the grandfather, father, and grandson were all addicted to alcohol. Given that the grandson was only five years old, Combe concludes that the "peculiar state of the organization . . . was in this case transmitted from one generation to another" (355).[32] Similarly, James Simpson claims that intemperance is shown to be an organic "affection of the brain" by "its being hereditary in families, and breaking out at the same age in several individuals of the same stock . . ." (10). Whether or not Brontë encountered such works, the novel clearly reflects the view that alcohol addiction, at least in extreme cases like Arthur's, has a heritable aspect. Brontë invokes hereditary concerns at both times the novel addresses Helen's decision to condition her son to hate wine by doctoring it with tartar emetic or forcing him to drink while ill. In justifying the practice in her diary, Helen states, "He was inordinately fond of [liquors] for so young a creature, and, remembering my unfortunate father as well as his, I dreaded the consequences of such a taste" (313). This comment corroborates Gilbert Markham's earlier aside that it was "generally believed that Mr. Lawrence's father had shortened his days by intemperance," and again reveals Helen's fear that her husband's tendencies, because they are innate, will reemerge in her son. Mr. Lawrence also cites hereditary transmission in defending Helen's tactics to Mr. Millward, claiming that extreme methods are called for "when a child may be naturally prone to intemperance—by the fault of its parents or ancestors" (66).

Helen's method of conditioning her son is a paradigmatic example of associationist psychology, but, significantly, it is a form of associationism that depends upon a physiological response. Unlike the other vices in which Arthur indulges, the consumption of alcohol involves an organic process and provides Helen with the rare opportunity to rewrite her son's physiological reaction to external influence. As she puts it, "I wish this aversion to be so deeply grounded *in his nature* that nothing in after life may be able to overcome it" (331, my emphasis). This is the sole instance in which Helen can successfully thwart her husband's counteractive influence in his own house because she can marshal the power of physiology against itself. This very tactic of using tartar emetic to associate drinking with nausea was also recommended by Caldwell precisely because phrenology proved the disease to be an innate physiological condition: "Drunkenness and its appetite are evils resulting no less from an organic cause, than an inflamed eye . . . if they be removed at all, it must be in conformity to organic laws." He notes that while "advice, remonstrance, warning, and denunciation [will] avail but little," inducing nausea by tartar emetic works due to the "effect produced sympathetically on the brain" (331).[33] The involuntary physical response to alcohol through the compound overrides, but does not eradicate, the inner compulsion to intemperance. Thus, although the therapeutic method seems to demonstrate the power of external influence to effect a dramatic change in behavior, it does not in any way create a change in the underlying tendency but only associates its arousal with an immediate negative response. Thus, Helen's tactic actually affirms the power of innate psychological drives to determine a person's behavior and actions.

As Gilbert points out to Helen, such a method does not inculcate virtue but simply removes the possibility of acting immorally: "What is it that constitutes virtue . . . ? Is it the circumstance of being able and willing to resist temptation; or that of having no temptations to resist?" (57). As Gilbert correctly deduces, the ethical implications of the practice touch on the subject of free will, since little Arthur will never freely choose to avoid alcohol but rather will recoil from it as he would from any unpleasant stimulus. Helen does not counter this assertion but does express regret that she cannot "render the incentives to every other

[vice] equally innoxious in his case" (57). Although the Markhams do not have the context to fully interpret her meaning, Helen discloses that her son's "case" is unique—a circumstance further underscored when she almost lets slip that her concern is owing to the character of her son's father: "[Why] not rather prepare for the worst, and suppose he will be like his—like the rest of mankind, unless I take care to prevent it?" (58). Following her ordeal with Arthur, innate dispositions and hereditary transmission now weigh heavily on Helen, so much so that she is far less concerned with the theological implications of her practices than with the practical results.

Helen's staunch defense of psychological conditioning at the beginning of the novel and near the end of the diary supplies a counterpoint to Brontë's presentation of the question of free will in the scene in which Arthur proudly displays his deficient moral organs. Helen's view at the beginning of her marriage was that regardless of the size of his mental organs, he still possessed the capacity to alter them through exercise. Helen's decision to use physiological association on her son comes quite late in her marriage, after she has become fully convinced that she cannot influence her husband, thus indicating an important shift in perspective on effecting psychological change. Whereas she earlier reasoned away Arthur's lack of "a proper organ of veneration" as a temporary circumstance, she now attends to biological signs of character with grave seriousness, even to the point of keeping the portrait of her husband that she initially wanted to destroy so that she can periodically "compare [her] son's features and countenance with [it] and thus be enabled to judge how much or how little he resembles his father" as he develops (191, 332). Given the novel's consistent interest in visible indices of innate character, Helen does not do this for curiosity's sake but in order to monitor and counteract the degree to which the father's biological inheritance manifests in the son.

Unfortunately for Helen, there is no physiological end run around Arthur's other vices, and the fact that she cannot effectively counteract her husband's psychological influence on her son is what finally authorizes her flight. In stark contrast to Helen's earlier insistence that her moral influence could save Arthur, by the end of her diary she invokes "influence" only to describe the danger her husband poses to her son: she wishes to take little Arthur where "he will be safe from [his father's] contaminating influence" (268), she wishes "to deliver [her] son from that contaminating influence" (280), and finally, she realizes that "it was absolutely necessary that [her son] should be delivered from his father's corrupting influence" (327). The reason Arthur's influence prevails over Helen's is because little Arthur already possesses "all the embryo vices a little child can show"—he is congenitally predisposed to develop the immoral character inherited from his father (298). It is only when Arthur is completely absent for months at a time that Helen makes any progress reclaiming her son, only to have her efforts quickly reversed when he returns (282). Ultimately, Helen realizes that the only effectual way to combat the impact of Arthur's biological and domestic influence is to completely control her son's environment by removing him from his father's home and carefully monitoring and regulating his interactions with the outside world. Or, as Helen puts it, "my child must not be abandoned to this corruption: better far that he should live in poverty and obscurity with a fugitive mother . . . the world's opinion and the feelings of my friends must be alike unheeded here . . . alike unable to deter me from my duty" (299–300). Although Helen breaks the law by absconding with her son, Brontë frames the act as one that adheres to a natural duty that materializes because she cannot change her husband and because she requires ideal conditions to foster her child's development.

As Helen explains to Arthur on his deathbed, her commitment to barring access to his son answers to a "higher duty" than obedience to him (364). Although her language marks this as a moral decision, the ethical aspect emerges from a hard-won pragmatism that recognizes both the futility of influencing her husband's character and the extreme susceptibility of her son to corrupting forces. Significantly, this moral position contradicts not only a father's rights under the law but also an ideological perspective on feminine influence that grounds itself in Christian values. What Anne Brontë superadds to her argument is physiological evidence of incorrigibility, thereby marshaling the rhetorical force of a specific scientific discourse against more nebulous understandings of women's supposedly inherent domesticating abilities. In doing so, the narrative effectively excises the compensatory aspect of feminine influence from a separate spheres ideology, which in turn exposes the way in which this model potentially imperils wives and children who have no recourse under the law.

Anne Brontë advanced her critique of the cultural myths surrounding a woman's role in marriage by linking it to a widely recognized scientific discourse of the day. What is important in tracing her use of phrenology is not simply to "decode" the language of a popular psychology but, rather, to pay attention to what she did with the foundational premise that human character was in important ways determined by the brain's material nature—a premise that has remained supported by all subsequent developments in neuroanatomy and physiological psychology. In a sustained way, the novel grapples with the social implications of the brain's materiality, highlighting the ways in which this premise contradicts or complicates mainstream values. This is perhaps the key difference between Anne and Charlotte's use of phrenology. For Charlotte, the science justifies Jane and Crimsworth's ascension on biological grounds, aligning their rise with the similar aspirations of the upwardly mobile middle class. Anne's application, however, works against the grain of traditional middle-class values, revealing the peril into which such cultural beliefs can place women. Perhaps more importantly, her particular use of biological essentialism brings into focus how the assumptions underpinning marriage laws are not universally just. Such an awareness provides us with an early instance of a woman writer harnessing the power of an appeal to biology to radically challenge the cultural assumptions that informed social relations.

Chapter 3

Harriet Martineau's Material Rebirth

In her last will and testament, Harriet Martineau makes an unusual request in the name of science: "it is my desire from an interest in the progress of scientific investigation that my skull should be given to Henry George Atkinson . . . and also my Brain if my death should take place within such distance of the said Henry George Atkinson's then present abode as to enable him to have it for purposes of scientific observation."[1] Atkinson, in addition to being one of Martineau's closest friends, was also the "ablest phrenologist" she knew, and in bequeathing him her skull and brain, she believed she would supply valuable evidence in support of the material basis of existence. Because she was a prolific intellectual, with a mind "well known" to the public, she reasoned that the structure of her brain could be more accurately correlated with its functions (*Autobiography* 1: 391). But even beyond its practical value to phrenological research, Martineau's bequest was, for her, a significant gesture that reflected some of her most cherished intellectual values as a materialist and agnostic: that the mind is the product of the brain (rather than the soul), that scientific research is crucial for the advancement of society, and that in death as in life what constitutes individuality is mere matter. As a symbolic act, the bequest also literalizes the separation of body and mind, privileging the latter and casting the former as ancillary.[2] The postmortem act would allow her to literally shed the body that had betrayed her so much in life as an invalid, and that which remained would be the most important part of herself: her brain, the origin and source of her individual identity and considerable intellectual labor.

For Martineau, the phrenological tenet that the brain was the explanation for, and origin of, all human behavior was clear biological evidence of what she had long sensed: "the *feeling* . . . that the theological belief of almost every body in the civilised world is baseless" (*Autobiography* 2: 283). Atkinson played a crucial role in Martineau's conversion from waning Unitarian to agnostic. As she puts it in her autobiography, although her "passage from theology to a more effectual philosophy was, in its early stages, entirely independent of Mr. Atkinson's influence," he was nevertheless the one who showed her "the shortest way round the corner" through phrenological theory (281). Martineau believed that phrenology offered empirical proof of her inner convictions, and she believed that its logic would convince others that naturalism, rather than theism, organizes the world.

Martineau attempted to share this message with the public in her *Letters on the Laws of Man's Nature and Development* (1851).[3] This strange work of epistolary prose, co-written with Atkinson, was the product of Martineau's continued interest in phrenology and mesmerism, but its central aim was to make an anti-theistic argument using science. In recent scholarship, the *Letters* have become a curious footnote in biographical criticism that attempts to maintain Martineau's literary reputation by deemphasizing the book's importance within the larger context of her work.[4] Martineau, however, emphasizes the book's significance in her *Autobiography* (1877), devoting an entire chapter to the conception, composition, and reception of the book, earning the *Letters* more autobiographical exposition than any of her other publications. In the chapter's concluding sentence, she explains that her estimation of the "great importance" of this work "may excuse, as well as account for, the length to which this chapter of [her] life has extended" (2: 370). For Martineau, this importance was both textual and biographical because the *Letters* served as her first public disavowal of theism. Martineau believed that in publishing the work she was jeopardizing her reputation as a public intellectual; she recalls, "It seemed to me probable that, after the plain-speaking of the Atkinson Letters, I might never be asked, or allowed, to utter myself again" (2: 343). The cursory scholarship on the *Letters* owes much to the assumption that it is primarily a book "about" phrenology and mesmerism. Martineau certainly respected both sciences and used them unapologetically in the *Letters*, but the work is really about physiological science, its connection to consciousness, and what this connection reveals about human identity. Thoroughly interdisciplinary in her interests, Martineau used the work to explore the possibilities of identity that revised the notion of what constituted the self. Her primary motive was to use phrenology's assertion of a physiological basis of mind to dislodge prevailing notions of identity from sociocultural concepts based on religious beliefs. To do so, she strategically appropriated the formal elements of confessional narrative to publicly perform a secular conversion, replacing the theme of spiritual growth and illumination with a personal exploration of self-knowledge through phrenology.

Among the women authors examined in this book, Martineau was the most explicitly feminist in both her work and activism. Throughout her life she wrote in support of women's employment, access to education, increased protections in marriage, and equal pay. She also played a crucial role in the fight to repeal the Contagious Diseases Acts, which denied women's right to liberty for the sake of men's health.[5] Yet, unlike the feminist phrenologists addressed in the first chapter, she did not base her arguments on the foundational premise that no biological distinction exists between men and women's brains. Nevertheless, in her essay "On Female Education," she asserts that "as long as the studies of children of both sexes continue the same, the progress they make is equal," which she finds "proof sufficient to [her] mind, that there is no natural deficiency of power . . ." (77). This statement is in line with Deborah Logan's observation in her biography of Martineau that "[a]s a feminist . . . Martineau demonstrated a visionary quality, often unappreciated in her time and in ours, aimed at eradicating the fundamental biological prejudice underlying all women's oppressions" (219). Insofar as the *Letters* foreground that there is a disconnect between what is known about the mind's innate power and its proper pursuits, the work has implications for women's equality in all the arenas Martineau elsewhere championed. While this was not her acknowledged aim in the text, the rhetorical untethering of the mind from a theistic origin of individual identity would, of necessity, demolish religious justifications for sexual inequality based on divine will. Rather than biology ratifying God's distinct purposes for the sexes, materialism nullifies God's existence in all religions, and with them the attendant traditions and social practices upon which those beliefs were founded. Further, her decision to communicate her break with theism through the confessional mode is of a piece with her tendency to adopt forms of writing traditionally associated with men. As Logan points out, the fact that the majority of Martineau's work is nonfiction is subversive in itself because the only legitimate genre open to women at the time was the novel (22–23, 224–25). Recognizing the *Letters* as confessional reveals that Martineau's infiltration of masculine literary territory extends from genre to mode because until the twentieth century, confessional writing, whether religious or secular, was almost entirely the province of men, from Augustine to Rousseau to de Quincey.

In addition to their subversive gender implications, the *Letters* represent an important phase in Martineau's private formulation and public expression of anti-theism. Biographical scholarship has tended to depict Martineau's "conversion" as one progressing from rationalist Unitarianism to Comtean positivism.[6] Valerie Pichanick, for instance, describes Martineau's acceptance of positivism as "at heart nothing more than" Unitarian doctrines "altered to conform to a new secular faith" (198). Viewing her conversion in this way, however, relies on a specific definition of the secular that fails to account for the philosophical sophistication of Martineau's position in the *Letters*, which ultimately argue that a correct understanding of the brain undermines the foundation of all existing institutions, Christianity included, thereby making possible what Martineau calls a new "study of Mind and Morals, and of much of external Nature" (95). Martineau's rhetorical position in the *Letters* thus makes explicit the conditions through which a radical break with inherited forms might occur by signaling and then replacing the theme of spiritual growth and illumination with a model of self-discovery based in biological science.

Further, in viewing the *Letters* in terms of Martineau's use of the confessional mode clarifies the work's specific function in relation to her autobiographical treatment of disbelief. In *Eastern Life, Present and Past* (1849), the *Letters*, and the *Autobiography*, Martineau addresses her break with Christian theism, but in ways that correspond to the generic conventions and rhetorical strategies that further her larger anti-theistic argument. As a historical study in comparative religion, *Eastern Life* made Martineau's theological dissent more or less explicit, but without including a personal admission of disbelief. As Maria Frawley observes, in *Eastern Life* Martineau "makes a direct correlation between scenery (outer life) and the ideas and emotions (inner life) that can be recovered from the landscape" (142). This thematic strategy afforded her both a tone of objectivity and the opportunity to "embrace recovery as a mode of inquiry" through history (135). While the scenery catalyzes reflection, the rhetorical presentation of Martineau's internalized movement away from theism through a travel narrative is of necessity externalized. In the *Autobiography*, however, Martineau offers a history of a different sort—one in which the history of the world does not catalyze a change within her, but rather, her personal narrative serves as a sign of future historical change. To this end, she applies a Comtean framework that progresses through theological and metaphysical phases to arrive at the final stage of scientific positivism. As Linda Peterson has argued, the "Comtean phases become Martineau's hermeneutic substitute for the biblical typology of orthodox spiritual autobiographers," allowing her to represent her "individual development as a specific version of general, human progress" (64). Unlike the observational travel narrative or the retrospective autobiography, confession is a performative externalization of personal knowledge in order to claim a new public identity. Whereas *Eastern Life* focuses on how external objects act upon the mind and the *Autobiography* attempts to demonstrate general principles about human progress through an individual's experience, the *Letters* publicly announce a privately apprehended truth about the origin and nature of cognition. For these very different aims—historical, philosophic, and scientific—Martineau uses (often irreverently and paradoxically) the conventions best suited to her larger ontological argument. Viewing the *Letters* in terms of Martineau's strategic use of the confessional discourse reveals the text to be integral to her ongoing examination of identity outside of faith.

Phrenological discourse was particularly well suited to this rhetorical aim. Caroline Roberts argues that phrenology appealed to Martineau because it explained "the relation between visible phenomena, or behaviour, and underlying structures, or causes . . ." (173).[7] Certainly, phrenology's apparent objectivity and emphasis on causation interested Martineau, who frequently made use of explanatory systems to articulate her perception of truths that were at odds with mainstream beliefs. In addition to its scientific status, however, phrenological discourse had also established and promulgated a way of speaking about secular conversion through the confessional mode. Autobiographical accounts by phrenology enthusiasts often cast the moment of accepting the truth of phrenology as one of immense personal significance. In a testimony typical of first-person narratives of phrenological conversion, one devotee recounts, "[w]hile I was unacquainted with the facts on which it is founded, I scoffed . . . and the pretensions of the new Philosophy of Mind, as promulgated by Dr. Gall, and now known by the term Phrenology. On hearing and conversing with his most eminent disciple, the lamented Spurzheim, the light broke in upon my mind . . . I became a zealous student of what I now perceive to be truth" (Mackenzie 7–8). In a similar testimonial, a convert explains, "Since commencing the study of Phrenology, a new light has dawned upon me, and various phenomena which were before perfectly inexplicable upon any known theory, are now of easy solution" (Macnish, "[Testimonial]" 15–16). Such narratives make use of the language that characterizes religious conversion in confessional prose—"testimony," "disciple," "new light"—and also follow the conventional arc of confessional autobiography by moving from a state of disbelief to a moment of sudden illumination that results in a life forever changed.

Phrenological conversion, however, was also frequently portrayed as offering benefits beyond intellectual illumination. A representative example appears in an account offered by a correspondent to *The Phrenological Journal*, who casts phrenological conversion as a form of professional salvation. Beginning with a description of his former life, the author explains that being "Born in the lap of luxury—bred in the tainted atmosphere of opinion" led to the "best years of [his] existence [being] passed in idle, if not in sinful pursuits." After becoming a military officer, his dissolute behavior increases until he finally decides to correct his character, and "the great instrument employed was phrenology" ("Remarkable Case" 342). After adopting a disciplined regimen of moral and intellectual mental exercises, he triumphantly emerges as a man who better knows himself and his capabilities. With his "new" character comes a new life more suited to his cultivated faculties: realizing that a soldier's advancement depends on the "number of victims" sacrificed for the country's cause, he "selected the more humble profession of the Civil Engineer, for which [he] believed, phrenologically, nature had made a fair provision." He concludes by noting that his life has since been prosperous, and that he hopes now to "aid the cause of that science" through the phrenological education of his own children and his public confession in the journal (343).

While the writer obviously makes use of the structure of Christian conversion narratives, his rebirth can hardly be classified as spiritual. His new life results from a change of occupation and a recognition of the practical benefits of a phrenological education. Anticipating the reader's possible objection to his substitution of the brain's materiality for an immaterial spirituality, he explains that only when "the principles of phrenology became developed" in him did "the beauties of Christianity become exalted" (343). Despite the respectful nod to Christianity, this narrative nevertheless reveals how the materialist philosophy following from phrenological theory implicitly challenged a traditional understanding of theism's relationship to personal identity. Rather than looking toward God to understand his purpose on earth, the author can only know God by first knowing himself, which can be facilitated only through an understanding of innate physiological attributes. Further, realizing God's benevolent design results not from an abdication of a quotidian life to follow Christ (characteristic of both the Pauline and Augustinian models of conversion), but rather from an acceptance of the professional life that suits his brain: a Christian conversion might yield a missionary, but a phrenological conversion produces a "Civil Engineer." By adopting the structure of a conversion narrative to recount his phrenological rebirth, the author elevates the importance of phrenology by aligning it with spiritual discovery.

In a similar way, Martineau used the imbrication of phrenological discourse and the confessional form to articulate her own conversion experience, the inciting incident of which was a miraculous withdrawal of the debilitating symptoms caused by an ovarian cyst.[8] In 1844, while living in Tynemouth, Martineau experienced immediate relief after a "laying on of hands" by the phreno-mesmerist Spencer T. Hall. As Susan Hoecker-Drysdale points out, this cure was one of the most significant and formative experiences of her life because it "demonstrated to her the validity of phrenology and the unity of mind and body. It showed the interrelation of being, feeling, and thinking. . . . It was most of all a successful empirical experiment; the results were tangible" (83). Like Atkinson, Hall studied the connection between mesmeric phenomena and phrenological theory, a subject variously referred to as "Phreno-Mesmerism," "Mesmero-Phrenology," or "Phreno-Magnetism." As Ilana Kurshan observes, phreno-mesmerism was "more of a practice than a theory," as it was limited to the stimulation of a mesmerized person's phrenological organs to produce a response demonstrating that organ's power (18). For instance, one could touch the organ of "tune" on a mesmerized subject's head and cause the person to sing. Atkinson describes his experience with the process, stating, "I found that I could excite into action any portion of the brain . . . I could play upon the head, and produce what actions I pleased, just as distinctly as you play upon the keys of the piano" (*Letters on the Laws of Man's Nature and Development* 39).[9] Like phrenology, mesmerism assumed a material cause for the effects it produced, positing the existence of a rarefied liquid permeating the body that led to mental discomfort or illness when unbalanced (Parssinen 105). Thus, phreno-mesmerists claimed to effect cures by absorbing physical pain directly from the sufferer, reasoning that since the entire body was controlled by the brain, drawing pain out of a specific organ would result in a corresponding cure elsewhere in the body.[10]

After her cure, Martineau became intensely interested in the experiential effects of mesmerism rather than with the phrenological theory of mind upon which the treatment was based. Only a few months after her experience with Hall, she published an account of her cure and the successful mesmeric treatment of others in "Six Letters on Mesmerism," which appeared in *The Athenaeum* in 1844 and was later republished as a book.[11] Martineau's interest in mesmerism has generated much recent scholarly interest in discussions of such Victorian issues as medical professionalization, femininity, and political economy.[12] Scant critical attention, however, has been given to Martineau's interest in phrenology, most likely because the *Letters on the Laws of Man's Nature* are viewed as an extension of her investigations into her mesmeric cure. Since the *Letters on Mesmerism* are not concerned with phrenology, Martineau's interest in the brain's organization and function understandably recedes into the background.

Although closely connected in theme, nearly seven years separate the first appearance of Martineau's mesmeric experience in print and the publication of the *Letters* in 1851. In the interim, she began her friendship and correspondence with Atkinson after their first meeting in 1845, and he explained the scientific thinking, both mesmeric and phrenological, that supposedly produced the salutary effects she experienced from treatment (Pichanick 183). Martineau demonstrates a serious interest in phrenology in a letter to Andrew Combe dated May 14, 1846. She writes:

> What I saw of mesmerism at Tynemouth taught me that Phrenology was true. . . . My intercourse with Mr. Atkinson has since presented the whole matter to me in a fresh & most interesting light. In him I see how much study,—how much time & earnestness,—how much sympathy & modest caution ought to go to make a wise phrenologist. . . . Whenever I come to Edinburgh . . . we may hope (may we not?) to have plenty of conversation on the two great affairs wh[ich] appear to be so closely connected,—Phrenology & Mesmerism. (*Collected Letters* 3: 58)

The letter reveals that Martineau had come to view mesmeric cures (what she "saw of Mesmerism at Tynemouth") as phenomena closely connected with a phrenological theory of mind. Further, it shows that Martineau's experience of cure preceded a complete understanding of, and respect for, the science that made her recovery possible. The gap between her cure and subsequent introduction to phrenology accounts for the divergent concerns in the related texts of the *Letters on Mesmerism* and the *Letters on the Laws of Man's Nature*, the former dealing solely with the experience of mesmeric trance, whereas the latter explores the origins of experiential phenomena as an effect of the brain's operations. In essence, the *Letters on Mesmerism* detail a conversion experience not fully grasped, while the more fully articulated statement of faith in scientific materialism emerges in the *Letters on the Laws of Man's Nature*: the first text records the miracle, and the second presents the creed.

By selecting phrenology as the science through which she would communicate her anti-theistic musings, Martineau participated in an ongoing public debate concerning the connection between phrenology and materialism. The charge of materialism and its atheistic valence not only influenced the lay public's opinion of phrenology but also caused dissention among phrenologists. The issue came to the forefront of phrenologists' attention in June 1842 at the fifth annual session of the Phrenological Association in London, which reflected preexisting tension between the London phrenologists, led by Dr. John Elliotson, and the Edinburgh phrenologists, led by George Combe. The former promoted phrenology as a discipline that could be reconciled with Christianity, while the latter camp began to publicly admit and endorse phrenology's implied materialism. At the meeting, the physician William Engledue delivered an opening address boldly asserting that mind has no existence independent of matter.[13] James Simpson then rose and made a motion to thank Engledue for his address with the qualifying statement that the comments concerning the solely material nature of mind should not be viewed as representative of the association's opinion. Simpson carefully constructed a third position in relation to the issue, claiming,

> [There is] not adequate proof to say, that, because we see only brain and its workings, there *is not* a power or energy beyond it. . . . On the other hand, when the immaterialist or spiritualist comes forward with his counter-assertion, that there *is* a power, or entity, beyond brain, called Spirit, which is not matter, he is equally unwarranted . . . in other words,

> we are utterly ignorant of the *essence* of mind. ("Report of the Proceedings of the Phrenological Association" 315, Simpson's emphasis)

In Simpson's formulation, the theory of the brain that phrenology sets forth exists outside the province of either religious or materialist doctrine. Reasoning that the brain's relationship to spirit may or may not reside outside the empirically observable, Simpson dismisses the necessity of engaging in the issue. Simpson's motion to thank Engledue with qualification was seconded, but a flurry of dissent erupted, many calling for the complete dismissal of Engledue's materialist position while others announced that they failed to see any reason for Simpson's qualifying remark (317–18).[14] Engledue's address generated even more intellectual heat in the following issue of the *Phrenological Journal*, which included a special section titled "Materialism and the Phrenological Association," printing no less than six letters responding to Engledue's claims. Five of these strongly disapprove of the address as an avowal of materialism, the most severe being that of John Isaac Hawkins, whose response succinctly summarizes the divide between materialists and anti-materialists within the association. After announcing his withdrawal from the organization "upon Christian principles," Hawkins states, "I would not separate myself upon slight grounds, especially as I continued a member for many years after numbers of my Christian friends had resigned in disgust at the Materialism so often obtruded on the meetings by influential characters" (50). In response to the confusion the entanglement of phrenology with materialism caused, he proposes forming a "Christian Phrenological Society" and related journal that would assert that "Christianity and Phrenology can be beneficially blended in investigating the condition of man" (51). Hawkins's vehement reaction illustrates the abiding prevalence of materialist notions among the phrenologists, which increasingly "obtruded" into the association's annual meetings, while simultaneously demonstrating how a materially based notion of mind proved problematic for theists.

Only one of the six letters, by T.S. Prideaux, overtly supports Engledue's stance by affirming the necessity of supporting a materialist view of mind if phrenology is to remain a science. Because legitimate sciences only investigate observable phenomena, he reasons that phrenology should follow suit by eliminating any consideration of God and concentrating on the brain's physiology:

the exhibition of feeling and intellect is invariably preceded by the development of a certain form and arrangement of nervous matter; and I fearlessly maintain . . . that the two former are the product of the latter. *Is it consistent with any sound principles of philosophy, gratuitously to burden science with an imaginary being, the existence of which is not demanded for the explanation of a single phenomenon?* I defy any one to answer this question in the affirmative. (54, Prideaux's emphasis)

Prideaux eschews overtly denying the existence of a spiritual nature or deity; however, his argument differs from the ameliorative third position proposed by Simpson by solidifying the legitimacy of phrenology's qualitative claims. Whereas Simpson argued that phrenology neither affirms nor denies materialism or spiritualism, Prideaux avers that the only knowledge of human nature available to science is through the "arrangement of nervous matter," and for this reason phrenologists must assume, without recourse to a deity, that an understanding of humans' moral and intellectual qualities are only physical effects of an aggregate of organs.

Like Prideaux and Engledue, Martineau belonged to this materialist school of thought, which she unapologetically champions in the *Letters*. Although the work addresses a broad range of interrelated issues, including positivism, the scientific method, and the nature of time, the unifying theme of the book is the material nature of mind.[15] In the first sentence of the first letter, Martineau inaugurates the epistolary discussion with an observation on Atkinson's unique insight into psychology: "I rather think the reason why we have so much pleasure in talking over, and writing about, the powers and action of men, and the characters of individuals, is, that your observations proceed upon some basis of real science . . ." (1). In his response, Atkinson asserts that the "different characters of men arise from the differences in the substance and form of their being. . . . For every effect, there is a sufficient cause; and all causes are material causes . . ." (7–8). Martineau's letter and Atkinson's response, respectively titled, "Inquiry for a Basis," and "Proposal of a Basis," imply not only that the material nature of mind forms the premise of the book but also that it serves as the foundation for all real knowledge. As Atkinson explains, because phrenological theory allows an investigator to "perceive precisely why men think as they do," anything one perceives with an attendant understanding of the natural basis of mind necessarily has more ideational substance and validity. Atkinson and Martineau envision the book's overarching purpose as a revelatory declaration that will ultimately free individuals from socially constructed knowledge by claiming identity as an effect of physiological organization. As Martineau asserts in the third letter, titled "Preparation of the Ground": "We deny, for our part, having any interior consciousness that informs us of any spiritual existence antagonistic to, or apart from, matter. If we once fancied we had, we have learned that it was through an ignorant and irreverent misapprehension of the powers and functions of matter" (14). Martineau argues that only by recognizing the soul as an illusory "shadow of ourselves" can humans progress to a further stage of existence founded on the secure knowledge that it is "matter through which Mind is manifested" (14–15).

Martineau and Atkinson advance this theory through an epistolary exchange that antagonistically parallels Christian doctrine to emphasize the radically different perspective that informs their scientific philosophy. Martineau replaces humanity's "misleading partiality" for origins with a knowledge of the brain, claiming, "By what combination of elements, or action of forces, I came to be what I am, does not at all touch my personal complacency, or interfere with my awe of the universe" (12, 13). Rather than identifying an immortal spirit as the seat of identity, she posits "physical fact" and "matter" as the most important sources of self-knowledge (12–13). Making the secular inversion of Christianity more explicit, Atkinson suggests creating a new "trinity" to "clear our understandings from the beginning," composed of "Matter, Form," and the immutable "laws of Nature" (169). Rather than humans being made in the image of God, he declares that the "material of the brain is an impress of Nature, and corresponds with the nature and principles of the world without" (193). In place of a Christian teleology involving "the belief of a future, and of a retribution" he proposes the "idea of the infinite and the eternal omnipresent Law, and principle of Nature" (189). Finally, and perhaps most shockingly for a nineteenth-century reader, he implicitly likens himself to Jesus: "A Christ for these times will be persecuted by these times; for a true prophet must ever be an offence to the world, until the philosophy of Man has become recognized as a true science, based wholly upon natural causes" (209).[16]

This pattern of appropriating conventional Christian forms to announce a new era of materialist thinking is rhetorically enacted in the work's sustained use of confessional narrative devices. The constituents of the conversion narrative are well recognized and typically include most or all of the following elements: seeing oneself in a public role, assuming a rhetorical posture of humility, awakening to formerly unknown truths, the recognition of a higher power's influence, and finally, the expectation of persecution for one's beliefs.[17] These thematic qualities are present in the writing of St. Paul and in the book of Acts, but they reemerge fully developed in Augustine's *Confessions*, the representative work of the confessional form. Each element also appears in the *Letters*, which Martineau more broadly casts as a confessional narrative in her *Autobiography*. Expressing her trepidation about personal disclosure, she reports that although she "anticipated excommunication from the world of literature, if not from society" for publishing the work, she and Atkinson were nevertheless obligated to share the good news: "[we] were both pursuers of truth, and were bound to render our homage openly and devoutly . . . we could not see them [members of the public] suffering as we had suffered without imparting to them our consolation and our joy" (2: 343–44). Although Martineau presents the publication of the *Letters* as a confessional act, it is admittedly confession with a difference. As Peter Brooks defines the concept, confession is "a speech-act that has a constative aspect (the sin or guilt confessed to) and a performative aspect (the performance of the act of confessing)" (52). Although the *Letters* have constative and performative aspects (through the creed espoused and the reception anticipated and experienced), the work differs from a conventional confession in that Martineau and Atkinson are not admitting to any "sin or guilt." What Martineau anticipates as the charge of sin rests solely outside of herself, in the mind of the public who will judge her admission. She performs confession not to be accepted by the social order (like the confessing prisoner or criminal facing the gallows) but to announce her willingness to be separated from it, like a martyr willing to be, as she puts it, "burned at the stake" (*Autobiography* 2: 346). Martineau views her own confession as a performative act that will, at the price of alienation, reconcile her to herself—her "freedom from old superstition" can be fully realized only if she articulates it to the world. As Foucault observes of the personal perception of modern confession:

> The obligation to confess is now relayed through so many different points, is so deeply ingrained in us . . . it seems to us that truth, lodged in our most secret nature, "demands" only to surface; that if it fails to do so, this is because a constraint holds it in place, the violence of a power weighs

> it down, and it can finally be articulated only at the price of a kind of liberation. Confession frees, but power reduces one to silence. (*History of Sexuality* 60)

In accordance with this model, Martineau perceived "power" as public opinion and her own disbelief in God and the immaterial nature of mind as a fundamental aspect of her identity after the experience at Tynemouth. As a result, Martineau believed she "had no further choice" than full disclosure, explaining, "any concealment would have been most imprudent. A life of hypocrisy was wholly impracticable to me, if it had been endurable in idea; and disclosure by bits, in mere conversation, could never have answered any other purpose than misleading my friends, and subjecting me to misconception" (*Autobiography* 2: 344). Before the work's publication, Martineau lives uncomfortably with herself, deeply committed to her personal beliefs while simultaneously aware of the unpopular reaction her avowal would occasion. In her formulation, this double existence subjugates her to power, as the price of silence is to be "subject" to "misconception." The act of confession reunites her public and private selves and wins her a sense of freedom through the cathartic surfacing of personal truth.

Although confession carries with it an unburdening of conscience, its public purpose lies in making known individual truths that correspond to universals. As Augustine explains in his *Confessions*: "This I desire to do [to confess], in my heart before you [God] in confession, but before many witnesses with my pen" (X.i). As M.H. Abrams observes, Augustine seeks self-knowledge through the act of confessing but simultaneously sees himself in a public role as one whose life has been transformed and thus must bear witness for the good of others (84). In the confessional mode, the personal becomes public so it may become personal again in the life of the reader, but this textual transference can occur only if that which the author communicates contains a truth beyond the personal. For Augustine, this greater truth is Christianity (the spiritual); for Martineau and Atkinson, it is the material nature of existence. The *Letters* attempt to expose as illusory the idea that individual freedom arises from spiritual development and posit in its place freedom from institutional dogma. Atkinson claims that modern man "remains ignorant of himself—of physiology and the laws of Man's nature," and thus "imbibe[s] a confused notion of unintelligible dogmas,—which are called religion, it is true, and which are vainly supposed to be all-sufficient to guide him through life . . ." (203). Since the "power to govern is in the knowledge of the nature of the thing governed" (286), and the nature of mind is material rather than spiritual, religion imprisons individuals in a false system of power.

In the work's preface, Martineau makes explicit her hope that by speaking plainly she and Atkinson might catalyze a secular awakening. Taking full responsibility for the decision to publish, she states that all systems of belief except that articulated in the *Letters* fail to satisfy her. Convinced of the validity of their views, she explains that personal conviction called for a public expression: "[It] seemed to me to require of us both the discharge of that great social duty,—to impart what we believe" (v). The "social duty" derives its importance from a readership of two kinds: those who are sympathetic but afraid to voice their approval, and those who will accept the truth but do not fully know the nature of that truth. She reasons that it "may be, or it may not be, that there are some who already hold our views, and many who are prepared for them, and needing them" (vi). Like Augustine, Martineau locates the importance of her public admission in hailing others who already share her beliefs. More important for Martineau, however, is the second kind of reader, who intuitively suspects that institutional forms have distorted perception yet has no basis for dismissing these constructed beliefs. Martineau claims that she and Atkinson are morally obligated to slake this thirst for enlightenment, explaining that having found "a spring in the desert, should we see the multitude wandering in desolation, and not show them our refreshment?" (344). Voicing their beliefs thus becomes the first act of obedience to the physiological basis of mind, and in so doing they make possible the secular, intellectual salvation of others who might not otherwise survive in the vast wilderness of ignorance.

The form Martineau selects to "discharge" her "great social duty," however, may seem to deemphasize her commitment. Atkinson's letters far exceed her own in length and anti-Christian content, and Martineau assumes a rhetorical position of deference in relation to Atkinson's opinions. In a most uncharacteristic tone, she prepares the way for Atkinson's responses by asking questions to which she already knows the answers. In her fifth letter, after expressing her dissatisfaction with other theories of mental philosophy, she remarks, "You will teach me better. You will open the matter to me . . . with some curious secrets that I know you hold thereupon. Now then,—what is our brain?" (28) She later positions herself as a willing student of phrenology, insisting, "you must give me some lessons from an actual skull, that may guide me in mesmerizing the sick," adding enthusiastically, "Now for the cerebrum! Where do you begin?" (72) Throughout, Martineau writes as if she were in need of constant instruction and correction, appealing to Atkinson's knowledge with giddy anticipation and unqualified admiration. Having achieved wide popularity and a considerable reputation as a writer, it seems strange that Martineau would willingly assume this posture in relation to a little-known phreno-mesmerist with only a handful of publications in the *Phrenological Journal* to his name.

Harriet Martineau's reverential attitude in relation to Atkinson in the *Letters* has been deprecated by critics since the nineteenth century. In his book-length response to the work, J. Stevenson Bushnan underscores this dynamic with the provocative title *Miss Martineau and Her Master* (1851), a phrase he uses throughout the treatise. Bushnan characterizes the relationship as one between an opportunist and his unwary victim, selected for her mental susceptibility: "Mr. Atkinson is her evil genius; he has suggested to her bewildered brain certain crude and distorted ideas. . . . To certain minds, owing . . . to peculiarity of structure, such ideas have a charm partly in virtue of their very crudeness" (11). A review in the *Zoist* similarly recognizes Martineau's worshipfulness as particularly deserving of ridicule, noting that "the point that especially gives its colour to the book . . . is the infantine simplicity with which the lady receives the decisions of her correspondent. No votary of Apollo ever travelled to the shrine of Delphi with feelings of more credulous submission than those with which Miss Martineau bows to the decrees of her atheistical Pontiff" (Anti-Glorioso 67). More recently, R.K. Webb has described Martineau's obsequious tone as "nothing short of embarrassing" (293), and Hoecker-Drysdale has characterized it as a lamentable pose: "This acclaimed popular educator, teacher of the public, assumes the role of student in sycophantic posture, playing the servile listener to the supposed expert. Quite out of character, [Martineau] assumes a position of awe and openness in the face of Atkinson's instruction on matters for which she, quite inaccurately, disclaimed all knowledge and opinion" (159).[18]

Martineau's rationale for deferential posturing, however, allows the *Letters* to involve a privileged listener—a quality that distinguishes confession from other generic modes of self-disclosure.[19] As Foucault observes, confession always "unfolds within a power relationship, for one does not confess without the presence (or virtual presence) of a partner who is not simply the interlocutor but the authority who requires the confession, prescribes and appreciates it . . ." (*History of Sexuality* 61). Within this relationship, the person to whom one confesses has the authority, occupying a position closer to the church (the priest), closer to the law (the interrogator), or closer to truth (God). As Peter Brooks explains, even as a means of self-exploration, in the confessional form the "expression of inwardness cannot, it appears, proceed without a responsive interlocutor in the search for self-knowledge" (95–96). In the *Confessions*, this "interlocutor" is God, who Augustine posits as the privileged "You." Only through God can knowledge come to Augustine, and any approach to God's understanding requires humility and deference in relation to this wisdom. In its legal, private, and religious forms, by engaging in the act of confession, the confessant implicitly admits lacking the power and knowledge already invested in the confessor and seeks legitimation through reciprocal understanding.

In the *Letters*, Martineau also strategically situates herself within a confessional power dynamic, but the knowledge before which she must humble herself cannot be mediated by someone representing conventional forms of authority. Like Augustine, who admits that he believes in a God he cannot yet fully understand, Martineau admits that she believes in, but fails to completely grasp, the material nature of mind. In her third letter, she includes a list of premises upon which she and Atkinson agree: a denial of an "interior consciousness" not connected with matter, the right to "require evidence" of those who would assert a spiritual existence (which "does not exist"), the belief that humans can "know only conditions," and an abjuration of the illusory "dreams" of an immaterial soul. Martineau presents the list in the form of a declaration of faith, prefacing each item with such inclusive pronouncements as "We deny," "We have a right," and "We abjure" (14). After establishing the core beliefs that she and Atkinson accept, she introduces the subject that forms the basis of inquiry in the *Letters*, what Atkinson can "teach" her: "If we cannot set ourselves back to the beginning of our reflective existence, and trace the whole course of our ideas and experience, you can teach me much of that particular department of matter through which Mind is manifested" (15). For Martineau, the origin and purpose of existence cannot be known, and for that reason it hardly warrants further interrogation; what can and should be investigated, however, is the "structure and functions of the brain" (27). Atkinson and Martineau argue that the need for a belief in a higher power roots itself in an egotistical desire to transcend nature's limits. An ideal object of desire, personified as God, satisfies the human longing for immortality and the assurance of being made in the image of an omnipotent and perfect being (206). According to Atkinson, the desire and its attendant posture of humility toward something greater than the self occurs naturally, but when individuals led astray by doctrine submit to a deity the "affections are perverted from their proper sphere." To remedy this, humans must "learn to *submit* to the rule of nature, and remember that unhappiness and discontent are selfishness—impossible to a truly heroic and loving nature" (207, my emphasis). Atkinson attempts to retain the human tendency to deferentially submit to a higher power but replace the object that commands obedience with nature's law.

In the twenty-third letter, this spirit of submission takes its fullest form as Martineau recalls her shame in once believing that even if God did not exist, no harm could come from believing that God served as "a model to Man,—the original of the image" (282). She realizes, however, that even a private acceptance of error causes her phrenological organs to regress in their development, "from the lowest faculties of perception up to the highest of conscientiousness, reverence and benevolence,—which ensues upon all tampering with our own best nature." Here, Martineau explicitly references three of the phrenological moral organs—"Conscientiousness," "Reverence" (also called "Veneration"), and "Benevolence"—positing the necessity of a physiological morality to ensure intellectual purity. When exercised toward the belief in a "lie" rather than toward the pursuit of truth, these organs cannot attain their highest development. While holding to error is an "incapacitating condition" aligned with "evil," faith in nature's laws improves the conditions of the human mind while promoting knowledge (283). In the rest of the letter she continues to employ the rhetoric of submission in relation to nature, advocating "obedience to Nature," "reliance on the immutability of Nature's laws," and a "sweet and joyful surrender to Nature" (283–85). In this way, she, like Atkinson, advocates retaining the form of religious devotion while replacing its object. Rather than moral growth through spiritual devotion and Christian learning, she calls for the improvement of "mental and moral health" through an acceptance of the material condition of natural causes and, most importantly, the cultivation of a spirit of unceasing inquiry into the structure and function of the brain. In this formulation, knowledge becomes the ideal, and humbly seeking it—first by acknowledging human error and ignorance concerning the mind's operation, and then dedicating oneself to scientific inquiry—becomes the moral pilgrimage required of all members in a progressive society.

In order to rhetorically model this attitude of reverence toward nature, Martineau enacts submission in her epistolary relationship with Atkinson. Claiming that he is the only person she has ever known to be "altogether above" prejudice, Martineau treats Atkinson not as an expert but as an exemplary figure embodying the proper attitude toward seeking true knowledge based in nature (12). Martineau can trust Atkinson's doctrine because it is not dogma but rather the tentative premises that emerge from a careful study of the only knowable first cause of consciousness: the brain. By reenacting through the *Letters* her own progress in studying the mind's material nature since first meeting Atkinson in 1845, Martineau provides the reader with a model of how one should approach novel ideas. She explains: "If we,—you with your habit of study, and I with my growing conception of what study is,—are daily sensible of the enjoyment of that 'perpetual spring of fresh ideas' . . . what must be the privilege of future generations who shall at the same time be more naturally free to learn, and find themselves in a bright noon-day region and season of inquiry!" (284). By exercising their faculties through the contemplation of the brain, the pair prefigures the freedom of thought that will be more fully experienced by future investigators. Through her self-conscious posturing as a disciple, Martineau models the appropriate habits of mind for one seeking enlightenment: a willingness to disavow preconceived ideas about the nature of the human condition, and receptivity to new ideas. Only by sweeping away the relics of inherited social, cultural, and religious assumptions can a "season" of knowledge begin.

After assuming the identity of a willing acolyte, Martineau further demonstrates the personal effects of new knowledge through another confessional theme: the experience of awakening characterized by emerging from darkness to embrace the new light of understanding. St. Paul's experience on the road to Damascus typifies the literal and symbolic movement into "light" through a conversion experience, which for Paul was precipitated by a "light from heaven" that flashed around him, rendering him blind. When Ananias restores his vision three days later, the former persecutor of Christians preaches about Christ in the synagogues, radically remade into an instrument of God for the conversion of the Gentiles (*New International Version Bible*, Acts 9: 3–20). In the *Confessions*, Augustine patterns his own mental experience after Paul's, casting his former confusion with darkness and his new clarity of conscience with light. In the garden at Milan, Augustine's conversion fittingly occurs after reading a passage by Paul (Rom. 13: 13–14) that rids him of disbelief "as if a light of relief from all anxiety flooded into [his] heart. All the shadows of doubt were dispelled" (153). Similarly, in Bunyan's *Pilgrim's Progress* (1684), the metaphorical movement of the wayward soul toward God's wisdom is a voyage toward light. Christian must advance toward "'yonder shining light,'" always keeping "'that light in [his] eye'" in order to obtain salvation (12).[20] The sudden apprehension of light allegorically represents the new or re-created world properly seen by the spiritually remade Christian wayfarer.[21]

Preserving the imagery of light and darkness and its metaphorical valence, Martineau communicates her own progress toward wisdom as one fundamentally counter to a Christian epistemology. The central tension in Augustine's *Confessions* is between two contradictory wills—one that is "of the body" and the things of this world, and the other that is "of the spirit," beyond crass matter and the senses (127). For Martineau, however, the movement toward enlightenment must progress in an entirely opposite direction. Because belief in the immaterial spirit has become so ubiquitously engrained in the concept of consciousness, it has achieved the position of an assumed truth. In her estimation, a genuine rebirth can occur only after a rejection of intangible spirit in favor of physical knowledge. Precisely because phrenology insists on the material nature of mind, the science strikes her as a useful tool for dislodging illusory notions about what constitutes human character. Remarking on the innovations brought to the science of mind through phrenological inquiry she states:

> The old field of (so-called) knowledge seems to melt away when we look into this new and tangible exhibition of the powers of Man: and in its stead spreads out a great unexplored region, with little in it clearly visible but the roads which are beginning to penetrate it here and there. We are passing out from the phantasmagoria to the dawn, where all is yet shadowy and solemn, but wherein the chief points are fixed, and we are sure of the East by the light that is in it. (95)

The "tangible exhibition" of mental power rests in the organs of mind that can be seen, felt, and subjected to scientific scrutiny. Martineau aligns this knowledge with "the dawn" and the "light" in the East, while shadowy apparitions or "phantasmagoria" bedizen past understandings of what constitutes human nature. Rather than understanding interiority through soul or a connection with the divine, she envisions understanding public and personal identity as a metaphorical voyage into the brain.

Martineau further claims that the persistent belief in the immateriality of mind has caused the history of mental philosophy to follow a false path. She explains in her opening letter that because "Natural Philosophy and Mental Philosophy are arbitrarily separated," science has progressed while understanding the mind remains stagnate in the quagmire of perpetual debate, stymied by the misrecognition of human existence as something containing an immaterial component (2). Metaphysicians "plead the totally different and incompatible nature of the two regions of inquiry, and therefore of the method of penetrating those regions," but Martineau argues that the two fields must be made one because understanding the mind through spirit cannot be systematized. When "spiritual agencies are at work" they can be "recognized only by each man for himself, by means of a special spiritual sense of which no one can give an account" (3). To illustrate the idiosyncratic views on the nature of self-knowledge, Martineau writes that on Sundays she amuses herself by asking everyone she meets, from the cowherd's wife to a poet to a Swedenborgian, what he or she thinks about the relationship between body and mind. The answers she receives affect her as if she were "carried back some thousands of years" when "men's instincts constituted the mythology under which they lived" (4). In her estimation, this false consciousness can be corrected by the light of science, which will render subjective musings on the soul the territory of objective and standardized investigation. She declares, "Now, Science has disabused us of our blinding and perplexing notions of spiritual anti-types of material things, and of spiritual interference in material operations; and we have arrived at the notion of chance-excluding LAW in the physical operations of the universe" (3–4). She later likens embracing science to the act of "coming out of a cave full of painted shadows" (219), aligning the state of blindness not with an inability to compass the immaterial nature of the soul, but rather with a preoccupation with spiritual, rather than material, reality. In doing so, she deploys the language of conversion against itself, casting supposed knowledge of the soul as intellectual darkness, and the rediscovery of the material conditions of the body and its power as a journey into light.

Enlightenment for enlightenment's sake, however, fails to provide the psychological benefits that attend religious conversion. Anticipating this, Martineau argues that a perceptual reorientation toward materialism can occasion experiential benefits similar to a mystical union with the divine, but she attributes these effects to as yet unexamined physiological properties of the brain. She avers that throughout history "all operation of one thing upon another, was concluded, before science existed, to imply spirit" (42). Although science had, by the nineteenth century, progressed far enough to prove that much naturally occurring phenomena resulted from physical laws, the brain's effects remained connected with the soul. Defining all instances of "spirit" as "Natural Magic which science shows, sooner or later, to be no magic at all," Martineau attributes the phenomena of oracles, prophecy, witchcraft, and visions to latent powers of mind that science has yet to explain. Believing that "whatever powers we have, have always been *there*," she asserts that finally "science is tracing [them] to their origin, or abiding place, in the brain" (123). Martineau's insistence on supposedly spiritual powers having "always been *there*" emphasizes not only that humans have consistently had access to extrasensory abilities but also that these seemingly otherworldly or miraculous gifts have their origin in an unusually well-developed organ in the brain. The effects of extraordinary mental operations may strike the individual experiencing them as the influence of a divine power, but Martineau attributes this misrecognition to the slow progress of scientific inquiry into the brain's structure.

In her thirteenth letter, Martineau offers her own testimony of the sensations caused by the brain's untapped extrasensory powers, describing the experience of exercising a "new faculty" (i.e., one yet unaccounted for by phrenologists) under phreno-mesmeric influence. "Nothing," she explains, "in the experience of my life can at all compare with that of seeing the melting away of the forms, aspects and arrangements under which we ordinarily view nature, and its fusion into the system of forces which is presented to the intellect in the magnetic state" (122). Martineau includes this testimony to discredit miraculous evidence of the supposed spiritual nature of mind. Her own encounter with what may have been termed a vision or mystical union with the divine only reconfirms her belief that the brain's capabilities have not been sufficiently explored. Rather than personifying and exteriorizing the "forces" felt, she directs her attention inward, to the brain's innate qualities. Reinterpreted through the lens of biological essentialism, Martineau uses her "visionary" experience to justify a return to, rather than an escape from, the body. Working against the Augustinian tradition in which the Christian sojourner must deny the pleasures of the body to secure inner peace through the spirit, Martineau claims that an escape from religious dogma produces pleasurable and healthful physical sensations. As she describes the effects of admitting openly her beliefs: "what a *feeling* it is,—that which grows up and pervades us when we have fairly returned to our obedience to Nature! What a healthful glow animates the faculties! What a serenity settles down upon the temper! One seems to have even a new set of nerves . . . no more pit-falls and rolling vapours,—no more raptures and agonies of selfish hope and fear,—but sober certainty of reliance on the immutability of Nature's laws" (283). Rather than positing the next world as a spiritual constant, Martineau locates stability in nature's laws. One receives compensation for giving up the "selfish hope" in an afterlife through a renewed vigor of the brain's organs—rebirth into a secular and materialistic worldview being accompanied not by a new life but by a "new set of nerves."[22]

Importantly, Martineau presents her rebirth as a "return" to nature, which gestures toward the confessional practice of distinguishing between secondary causes and a primary or First Cause. Augustine demonstrates this key aspect of spiritual autobiography in his continual reinterpretation of experience through the lens of a providential plan.[23] Martineau, however, views the ideational connection between secondary causes and a divine will as absurd: "To think that they [theists] begin with the superstition of supposing a God . . . who is their friend and in sympathy with them, and the director of all the events of their lives, and the thoughts of their minds. . . !" (218) Nonetheless, she and Atkinson retain the notion of a force greater than the self through natural law, the impetus that truly directs the operation of the world. She defines this power as "a system of ever-working forces, producing forms, uniform in certain lines and largely various in the whole, and all under the operation of immutable Law" (219). Martineau argues that the rules of matter's physical manifestations can be systematized and scientifically understood, but the refusal to apprehend this law and its influence has left individuals estranged from a true knowledge of their unconscious motivation.

For Atkinson and Martineau, self-knowledge built on the premise of a divine plan results in a false notion of individual identity. For Augustine, understanding one's purpose follows from reading scripture and engaging in prayer, activities that lead the individual toward the true self, created by God for a specific spiritual purpose. Opposing this view, Martineau and Atkinson attribute unconscious motivation to the directing force of matter that operates in accordance with natural law. As Atkinson states of the boon afforded by a phrenological understanding of the brain, "Man, who has seen himself only under a mask, will at length see himself as he really is. This will be the greatest, and the only wholly true and lasting revelation in regard to Man, the world has received" (190). In his formulation, the reflection of human character has been obscured by the artificial "mask" of spirituality. For individuals to truly "see" themselves, they must first frankly confront their identity as material phenomena influenced by the impersonal law of nature.

Phrenology served this purpose for Martineau by supplying her with a tenable theory of individual identity. As she explains in the *Letters*, long before her association with Atkinson she had struggled in vain to find some way to account for human consciousness. In her youth, she accepted Associationism as the most rational theory but ultimately found it too limited (118–19). As her mesmeric experience of a "new faculty" demonstrated, however, some aspects of mind cannot be attributed to experience. While one's personal development might be influenced by environmental factors, "Man's Nature" is organic and innate.[24] Martineau attributes the reason individuals do not already know themselves to a general lack of scientific knowledge about the brain. She explains that in her own search to "understand [man's] nature, and his place, business and pleasure in the universe," she floundered in ignorance: "it is strange to think how many books I have read, and how often over, and what an amount of hours I have spent in thinking, and how many hundreds of human beings I have watched and speculated upon, without being ever, for one moment, satisfied that I knew what I was about,— for want . . . of some scientific basis for the inquiry, and of some laws manifesting themselves in its course . . ." (2). Understanding the mind to have a material nature that operates through natural laws refigures consciousness as a state of *being* rather than something in the process of *becoming* through experience. Martineau and Atkinson view achieving self-knowledge as a process—that is, they distinguish between the self that is and the self that one consciously knows—but coming to realize this knowledge requires a voyage inward rather than outward, into the tissue and fibers of the brain rather than an external relationship with God. As Martineau states, "I am what I am;—something far beyond my own power of analysis and comprehension" with "an origin in matter which cannot think, and forces which cannot feel" (13).

If the text of the *Letters* itself operates as the constative element of Martineau's confession, the performative aspect appears in Martineau's expectations regarding the book's reception and her hyperbolic defensiveness in reaction to her critics. In the nineteenth letter, she boldly announces not only her atheism but also the contempt she has for theism:

> To me . . . it seems absolutely necessary, as well as the greatest possible relief, to come to a plain understanding with myself about it: and deep and sweet is the repose of having done so. There is no theory of a God, of an author of Nature, of an origin of the universe, which is not utterly repugnant to my faculties; which is not (to my feelings) so irreverent as to make me blush; so misleading as to make me mourn. (217)

Unsurprisingly, this bold avowal of atheism shocked and dismayed Martineau's contemporaries. Charlotte Brontë wrote of the *Letters*, "It is the first exposition of avowed atheism and materialism I have ever read; the first unequivocal declaration of disbelief in the existence of a God or a future life I have ever seen. . . . If this be Truth, man or woman who beholds her can but curse the day he or she was born" (qtd. in Gaskell, 441). Patrick Brontë expressed a similar opinion when he wrote to Martineau on November 5, 1857, observing, "your unfortunate book on Atheism, made you many opponents and enemies, and gave a smack to those who gave you credit for reasoning powers." News of the book's negative reception reached Edinburgh soon after its publication, prompting George Combe to write Atkinson on October 19, 1851, stating, "I wholly dissent from the obloquy that has been thrown on you & Miss Martineau, for the publication of your letters." Even complete strangers were prompted to respond: a Mrs. Harcourt, anticipating Martineau's imminent death and certain damnation, sent two personally annotated Christian tracts accompanied by a letter begging the perusal of both and "hoping that the facts there stated may be made by the blessing of God the means of rousing you from your present fancied security." Of more personal significance was James Martineau's negative criticism, which appeared in his vitriolic review of the book in the *Prospective Review*, titled "Mesmeric Atheism." James took umbrage at the authors' insistence that free will did not exist, the denial of God as a First Cause, and especially their disbelief in a spiritual nature of mind. He writes, "So far as Science has effected the 'exorcism of spirit' from nature, has science produced, we believe, only delusion" (257). James approached the *Letters* from a viewpoint diametrically opposed to the doctrine espoused by Martineau and Atkinson, arguing that what they found illusory (spirit) was truth, and what they claimed to be true about the mind (materiality) was delusional. Point by point, James's review reified the very position the pair had attempted to dismantle.

Although the public reaction was largely negative, Martineau's outrage over the outcry seems exaggerated given the unorthodox nature of the views she so frankly expressed and the fact that she fully anticipated, and even braced herself for, an unfavorable reception. On November 6, 1850, before submitting the manuscript to Edward Moxon for publication, she warned him that it was "daring to the last degree; & the public wh[ich] certainly *is* ready for such works, may not be *your* public" (*Collected Letters* 3: 174).[25] Despite being steeled for the uproar, and her oft-reiterated belief that it is "never worth while . . . to notice reviews" (*Autobiography* 2: 354),[26] Martineau appears to have noticed, with great particularity and interest, the negative reception of the volume. In her *Autobiography*, Martineau discusses in detail the persecution she expected and received in relation to the book's publication. After moving to Ambleside, she avoided visiting her neighbors, knowing that "a book was in the press which would make them gnash their teeth" (2: 350). At length she describes the reactions of two unnamed "intimate friends" who, after the *Letters* appeared, found Martineau's beliefs "too much for them" and "blamed [her] excessively" (2: 352–53). One of the two women "went about every where, eloquently bemoaning [Martineau's] act, as a sort of fall, and doing [her] more mischief . . . than any enemy could have done . . ." (2: 353).[27] She also writes of neighbors who stopped speaking to her, of a decrease in visitors, a "diminution of letters" in general, while "scolding letters" from friends multiplied (2: 355–57). She claims that had she not been prepared for the reaction and fortified by Atkinson's support, she "might have declined into a state of suspicion, and practice of searching into people's opinion" during the ordeal (2: 358).

This emphasis on her suffering, however, also fits within the framework of the confessional arc. In order to perform confession, the convert must assume a posture of righteous fortitude characterized by an expectation of persecution and a willingness to accept censure. Martineau casts her experience of the hostile reception in precisely this way:

what did it matter whether people who were nothing to me had smiled or frowned as I passed them in the village in the morning? When I experienced the still new joy of feeling myself to be a portion of the universe, resting on the security of its everlasting laws . . . how could it matter to me that the adherents of a decaying mythology . . . were fiercely clinging to their Man-God, their scheme of salvation, their reward and punishment, their arrogance, their selfishness. . . . To the emancipated, it is a small matter that those who remain imprisoned are shocked at the daring which goes forth into the sunshine. (*Autobiography* 2: 355–6)

What Martineau claims did not "matter" to her personally in fact mattered very much in her public performance of persecution. Fittingly, she refers to the volume as her "testimony" (2: 348, 370), compares herself to Martin Luther (2: 351–52), and includes an anecdote in which a friend compared her to St. Paul (2: 353). The motive behind this posturing as a martyr surfaces when she writes, "I am endeavouring now to revive the faded impressions of any painful social consequences which followed the publication of the 'Atkinson Letters,' that I may not appear to convey that there is no fine to pay for the privilege of free utterance" (2: 358). Martineau, it appears, *wanted* to pay a price, and by so paying join the ranks of other reformers persecuted for their beliefs.

In carefully detailing the degree to which she was ostracized for her phrenologically inspired beliefs, Martineau made use of the same persecution rhetoric assumed by phrenologists in defense of their science. Although phrenology was the most popular science among the masses, many phrenologists remained indignant at having failed to gain acceptance among the scientific elite. *The Phrenological Journal* frequently included articles bemoaning such treatment as a form of unjust censure.[28] One of the most elaborate examples is George Combe's lengthy rebuttal of Francis Jeffrey's critique of phrenology in the *Edinburgh Review*, which casts phrenologists as "converts" who have "faith" in their science (6) and have "openly and audaciously professed [their] belief" in the face of overwhelming prejudice (2). According to Roger Cooter, adopting the pose of persecution was crucial to phrenology's rise because it allowed phrenologists to consolidate their identity in opposition to antiphrenologists, and because it helped to recast the science's marginalization as the effect of close-mindedness rather than a lack of legitimacy (*Cultural*

Meaning 82–83). Thus, the use of persecution rhetoric in phrenological discourse was more strategic than sincere in that it allowed phrenologists to assume a distinct public identity as martyrs for the cause.

In a similar way, Martineau's response to unfavorable criticism repeatedly and insistently ties the injustice of her treatment to her resolute moral allegiance to an intellectual conviction, thereby demonstrating that her persecution serves a principled purpose by allowing her to translate an avowal of belief into a new identity. In a letter to Helen Martineau on July 11, 1851, Martineau explains why she completely severed her relationship with James, listing his insults to Atkinson, the abuse of brotherly privilege, and the public nature of the attack as possible reasons for offense. She then, however, emphasizes that the decision to "disown" him had nothing to do with personal considerations, explaining, "I will not describe to his wife the qualities by which he would, without that public consideration, & apart from my friendship with Mr Atkinson, have compelled me, after the publication of that review, to disassociate myself from James, as unworthy of my esteem. . . . It is not personal offence, but moral reprobation that I feel" (*Collected Letters* 3: 203). Martineau emphasizes that despite James's lack of brotherly consideration, the real reason she must separate from him is due solely to her inward adoption of a principled moral stance. Similarly, in a letter to Arthur Nicholls on November 10, 1857, regarding Patrick Brontë's willingness to go "out of his way" to insult the *Letters*, she concludes with the observation: "It is no pleasure to me to see clergymen showing such a state of temper & of conscience. . . . If such conduct had been exhibited by two priests in the diocese of Tuam, you would have known how to estimate & characterize it" (*Collected Letters* 4:49).[29]

As an experienced public intellectual who, of her own admission, knew better than to notice criticism, Martineau's disproportionate reaction to the outcry seems, much like her posturing as Atkinson's pupil, completely uncharacteristic. Yet, it is entirely consistent with confessional discourse: in order for confession to be recognized as such, what one claims to believe must be attended by a testimonial act that legitimizes its signifying power through a public performance. In this vein, Atkinson repeatedly emphasizes in the *Letters* that a price would, and must, be paid for their admission, noting, "the history of Man is a history of the persecutions of the world's benefactors" (171). Later he repeats the sentiment, observing, "We may preach these things, and men will think us mad, or something worse. Truth and the progress of humanity have at every step of advance been thought a desecration . . ." (190). Nevertheless, Atkinson knew that as a public figure Martineau had a reputation to lose, and would inevitably bear the brunt of the burden. In a letter Martineau received from Atkinson in the aftermath of publication, he writes:

> You have no time to be miserable and repent,—have you? [N]o time to be thinking of your reputation or your soul. Your cheerful front to the storm and active exertions will make you respected; and remember, the Cause requires it. It would be hard for a Christian to be brave and cheerful in a Mahomedan country, with any amount of pitying and abusing; and so you have not a fair chance of the effect of your faith on your happiness in life,—as it will be for all when the community think as you do, and each supports each, and sympathy abounds. (qtd. in *Autobiography*, 2: 361)

Atkinson and Martineau's understanding of what the "Cause" required extends the significance of their work beyond its written message. Although words may be the "leading incident" in performing an act, they usually are not (if ever) the only thing required, and the pair well understood that an act must accompany their statement of belief in order for those words to "do" something (Austin 8). As Atkinson makes clear, the way in which Martineau weathers the "storm" of the public outcry will also clear a path for future materialists. In other words, her speech becomes an act *only if* she pays a price. As Atkinson observes in another letter, "With regard to what they say about us. . . . It seems that we ought to have something to bear" (*Autobiography* 2: 365).

Martineau believed that her secular conversion was the most important and formative event in her life, and she conceived of her belief in the material nature of mind as the foundation of her mature philosophic outlook. In a letter to G.J. Holyoake written on February 15, 1855, about whether declining health would allow her to write all or only portions of her memoir, she observes, "no one but myself can properly do the most important part,—the true account of my conscious transition from the Xn faith to my present philosophy" (*Collected Letters* 3: 350). Ultimately, however, Martineau's philosophy precluded the demonstrative aspect of faith conventionally associated with conversion. Paul, Augustine, and St. Anthony displayed their belief through a radical change in character, a personal change consonant with "the crises, cataclysms, and right-angled changes of the Christian pattern of history" (Abrams 48). While discontinuity and rupture with the old self is entirely possible in the Christian model, Martineau's belief that "it is enough that I am what I am" could not illustrate inner change through a radically different mode of existence. Ultimately, her doctrine is not a rebirth into a new life but a rebirth into an acceptance of an essential and inherent character—a transition away from the belief that there was ever a "new life" to be had. Something else was needed to make her confession an act, and she found the means in assuming Atkinson's projected role for her as a freethinking martyr who would rejoice in her persecution for the cause of materialism.

Throughout the *Letters*, Martineau attempted to envision and embody, to write and to become, the model of a new kind of person—one who would not locate consciousness outside the mind but within its someday knowable mysteries. The work is not persuasive but apocalyptic: Martineau's attempt to sound the death knell of spirit. As enigmatic as the *Letters* may seem, the work was prescient in its assumptions about how subjectivity would be viewed in the future. The ideas that brain is the organ of mind, that personality is in large part owing to our physiological dispositions, and that localized mental functions account for behavior are now mainstays of neuroscience and psychology. The work's legacy might be more accurately characterized as a prophetic announcement of things to come, appropriately recognized in its day and ours by James Anthony Froude's characterization: "Such a book as this is a strange echo of . . . forebodings. We may turn away from it, affect a horror of it, slight it, laugh at it; but it is a symptom of a state of things, it is the first flame of a smouldering feeling now first gaining air, and neither its writers, nor we, nor any one, well know how large material of combustion there may be lying about ready to kindle" (433). Martineau was among the first to set foot upon the disputed territory of a new "state of things," risking her public reputation for her personal belief in science's ability to revolutionize the way we view ourselves in the coming age.

Chapter 4

Mary Elizabeth Braddon's Physiological Critique of Social Identity

In 1867, Mary Elizabeth Braddon reeled from the blows of a critical attack in *Blackwood's Magazine*. The article, attributed to Margaret Oliphant (although Braddon always assumed it was written by a man), takes aim at women sensation authors in general, and Braddon in particular, for penning female characters who display "sensual passion" and an "appreciation of flesh and blood" (259). Oliphant speculates that this misrepresentation of polite English womanhood was owing to the fact that Braddon herself "might not be aware how young women of good blood and good training feel" (260). She then goes on to make charges of plagiarism along with thinly veiled allusions to the impropriety of Braddon's private life. According to one biographer, the article was "the most painful critical blow she had ever received" (Wolff, *Sensational* 203).

Braddon's immediate response to this provocation was to write to Edward Bulwer-Lytton and request his influence with the magazine in persuading the article's author to correct the inaccurate claim that her novels contained "the lurking poison of sensuality" (143). In defending herself against the charge, she first offers the "evidence of [her] books," and then the evidence of her cerebral development: "Of all horrors sensuality is that from which I shrink with the most utter abhorrence—and to you, Lord Lytton, as a phrenologist, I may venture to say—without fear of provoking ridicule—that all those who have examined my head phrenologically know that this sin is one utterly foreign to my organization, that indeed, the great weakness of my brain is the want of that animal power" (qtd. in Wolff, 143). Braddon's self-defense makes use of two distinct types of evidence: the documentary record of what she had already written, and a biological index of what she is capable of doing given the structure of her brain. Whereas the former argument only offers proof of what she has already done, her innate nature goes further by showing that she lacks the "animal power" necessary to depict female sensuality in her novels in the first place. For Braddon, a former actress then living with a married man, an argument about essential identity that could be completely detached from public opinion offered an attractive mode of self-justification: despite what people like Oliphant might think about the things she had done, she was not the person they thought she was, and her brain proved it.

The discrepancy Braddon identifies between phrenological knowledge and public opinion neatly foregrounds the dynamics that underpin a central issue in her early fiction: the relationship between a correct understanding of human psychology and the power to render that knowledge meaningful within larger systems of understanding. In both *The Trail of the Serpent* and *The Doctor's Wife*, she posits that the brain's materiality can contradict and invalidate the assumptions that underlie social formations. At the same time, these novels recognize that such knowledge is often powerless to alter the circumstances of those it most affects. While within the narratives, the gulf separating power and knowledge emerges as frustratingly unbridgeable, the novels themselves work to close that gap through a cultural critique that equates truth with a correct reading of innate identity. In *The Trail of the Serpent*, this critique operates by aligning the reader's perspective with that of a series of knowing but powerless individuals whose central task is to translate physiological knowledge into a form of truth recognized by institutional forms of power. Using a similar strategy to make a far more pointed feminist argument in her realist novel *The Doctor's Wife* (1864), Braddon examines the incommensurate relationship between a woman's extensive mental capabilities and the meager professional and personal opportunities available to her. By pitting phrenological assessments against the logic underpinning social roles, Braddon's work exposes the blind spots in ideological systems based in long-held assumptions about human—and, in *The Doctor's Wife*, women's—character. Nature, in other words, exposes the cracks in the faulty and unjust construction of the social world.

Up to a point, Michel Foucault's formulation of the interdependent relationship between knowledge systems and power clarifies the conflicts between competing truth claims in these novels. Foucault locates a discourse's ability to appear as true as an effect of its alignment with institutionally sanctioned systems of power that, in turn, are legitimized by the truth such power authorizes. From such a perspective, phrenology, or any form of physiological psychology, is a system of knowledge like any other that reveals in its rise and fall a changing relation to power. Insofar as Braddon fictionally *deploys* physiological knowledge, however, it is radically opposed to institutional forms of power and operates as a critique of the codified processes that invalidate or work to eliminate competing truth claims. Understood as offering a challenge to established systems through an appeal to essentialism, these novels participate in "detaching the power of truth from the forms of hegemony, social, economic and cultural, within which it operates" even as the forms of knowledge they champion remain within a larger system of power (Foucault, *Power/Knowledge* 133).

First serialized in 1860 as *Three Times Dead*, *The Trail of the Serpent* is Mary Elizabeth Braddon's first novel, and arguably the first work of British detective fiction.[1] Although the serialized version of the novel was not a financial success, the revised version reissued by John Maxwell sold a thousand copies in its first week (Wolff, *Sensational* 99). Although long overshadowed by *Lady Audley's Secret* and *Aurora Floyd*, *Trail* has recently generated considerable critical interest for its positive depiction of a disabled detective, its ambiguous relation to genre, and its representation of contemporary social issues.[2] The novel is a blend of sensation and detective fiction, combining the shocking scenes and subject matter of the former with the latter's interest in the investigative process. Unlike much sensation fiction, however, the novel does not conceal a secret from the reader, as the narrator reveals the identity and intentions of the criminal almost at the outset, and while much has been said about the novel's adherence to the conventions of these genres, far less attention has been paid to this singular quality that crucially structures the suspense of the narrative.[3]

The novel follows the criminal career of Jabez North, the detective work of Joseph Peters, and the plight of the falsely accused Richard Marwood. Dissatisfied with his life as a schoolmaster, North plots to murder and rob the wealthy fortune-hunter Montague Harding, who has recently returned from India to visit his widowed sister and her scapegrace son, Richard. After securing Richard's promise to reform, Harding agrees to make his nephew his heir and also gives him all his ready money to help establish him in a new life. Later the same night, North breaks in and murders Harding but fails to find the money he expected because it is in Richard's purse. The next day, two members of the Slopperton police, including the mute detective Joseph Peters, apprehend Richard and charge him with the murder. Peters, however, instantly recognizes Richard's innocence by closely reading his physical reaction to the accusation. Later the same day, Peters encounters North at a pub, and when he surprises him with a tap on the shoulder, finds in his physiological response "the very same look that [he] *missed*" in Richard's face, and realizes that he is the true murderer (245). Secretly working against the police and prosecution, Peters councils Richard to feign madness at the trial. After being found not guilty by reason of insanity, Richard is removed to the local asylum where he remains for eight years. Meanwhile, North forges checks stolen from his master, assumes a false identity, and departs for the continent. Eight years later, Richard escapes from the asylum with the help of the now independent Peters, who Richard's mother has engaged exclusively as a private investigator. Soon after Richard's escape, North returns to London after a successful criminal career abroad under the alias Raymond Marolles. Peters, aided by Richard's friends, finally succeeds in apprehending the true murderer and clearing Richard's name.

Braddon's critique of institutional power in the novel is wide ranging, taking aim at the police, the law, and the asylum. As a detective novel, the suspense inheres in the gap created between the known and the unknown—that is, by setting up the expectation of resolution and then delaying it. Unlike much later detective fiction, however, it does not withhold the identity of the criminal but, rather, how he will be proven guilty. As one critic puts it, the novel "is not so much a whodunnit as a 'howdunnit,'" because the "interest lies in the process of detection: how will enough evidence be found to convince the police to bring North to justice?" (Willis 409). While this generic description certainly accounts for the "knowing" position of the reader, it somewhat mischaracterizes the actual method of detecting the criminal: Peters does not identify North through the accumulation of circumstantial evidence but through the observation of physiological signs of guilt and deviance. Further, throughout the novel, Braddon equates possessing correct knowledge of human character with the position of the outsider: a phrenologist denounced as an impostor, a mute detective disregarded by the police, a physiologist ignored by the scientific establishment for his interest in the occult. What all of these figures also share is a method of detection based in reading biological and physiological signs of character. While North can successfully lie and easily deceive, his body betrays him—only nature escapes the system he has learned to game.

This truth, however, has no value within the respective systems it circulates because the social structures with power (the police, the law, the town itself) refuse to acknowledge the biological premise that ratifies it. Because the omniscient narrator reveals early on that North is the murderer, the narrative corroborates the conclusions of these marginal figures and validates their methods against established procedures. In this way, the narrative structure that inaugurates and sustains the novel's suspense relies on a disjunction between two rival epistemologies and an unequal distribution of power that is inversely related to the degree of correctness of the competing truth claims. The physiological truth the narrative endorses has no power to change the characters' circumstances, while the form of "truth" recognized by the justice system and the court of public opinion has the power to incarcerate Richard and remove North from suspicion. This arrangement creates an oppositional relationship between those who can read character rightly (the narrator, reader, and the physiological detectives) and those who do not (the Board of the Slopperton Union, the police, the law). The "gap" that creates and sustains the narrative's suspense, in other words, crucially depends on the separation of the actual knowing subject from power, and this gap *only* emerges because these figures can detect innate character. Institutional apparatuses and the epistemologies that support them are the ultimate antagonistic forces in the narrative insofar as they operate as the central impediments to the plot's resolution.

The phrenological encounters that bookend the novel are emblematic of the forces underpinning this conflict. In the opening pages of the novel, an itinerant phrenologist delivers an ominous public reading of Jabez North's remarkable skull. Contradicting the town's belief that North is a pious young man, the phrenologist reports he had "never met with a parallel case of deficiency in the entire moral region, except in the skull of a very distinguished criminal, who invited a friend to dinner and murdered him on the kitchen stairs." Outraged, the townspeople dismiss him as "an impostor, and his art a piece of charlatanism . . ." (7). At the end of the novel, however, the phrenologist triumphantly returns with other phrenologists in tow, all eager to take casts of the murderer's extraordinary skull. The narrator reminds readers that the original phrenologist "had given an opinion on his cerebral development ten years before, when Mr. Jabez North was considered a model of all Sloppertonian virtues and graces." Although he was then "treated with ignominy for that very opinion," he is "now in the highest spirits, and introduced the whole story into a series of lectures, which were afterwards very popular" (396).

Braddon's decision to open and close her novel with a phrenologist's persecution and eventual vindication is significant for narrative and thematic reasons. In terms of the narrative, the phrenologist's assessment of North is the first indication that he is not what he seems, creating the conditions necessary for suspense. Thematically, it signals the possible inaccuracy of social codes that render individuals legible within a community. Underlying the initial conflict between town and phrenologist rest two opposing assumptions about the origins of human character and how it might be accurately assessed. While the phrenologist bases his reading on physiological signs of an innate psychology, the townspeople anchor their opinion in history, observation, and communal pride. Discovered by a river as an infant, North was cared for by a variety of charitable societies and local institutions; and so "Slopperton," the narrator explains, "had in a manner created, clothed, and fed him . . . reared him under the shadow of Sloppertonian wings, to be the good and worthy individual he was" (7). The phrenologist's suggestion that North might have murderous tendencies offends the community because it contradicts the efficacy of collective environmental influence and the self-reflexive ethos that supposedly shapes not merely Jabez's personality but what kind of knowledge counts as an accurate index of "truth" within the communal arrangements that organize this society. The town's final acceptance of this method of assessing character illustrates that a dramatic shift has occurred—one that values "cerebral development" over reputation and the claims of nature over social influence.

Braddon's decision to conflate criminal behavior with physiological predispositions, particularly considering her use of phrenology to bookend the novel, was likely modeled on popular phrenological case studies of actual criminals.[4] Criminals were of continual interest to phrenologists as they constituted, in the words of one enthusiastic collector of skulls, "almost a type of character in themselves" (Deville, "Account" 20).[5] Phrenologists routinely attended public executions to sketch the profiles of the condemned or to take casts from the heads of recently executed criminals. Phrenological books and journals often included narratives of the lives of criminals that drew correspondences between predisposing organs and the unlawful acts they supposedly precipitated.[6] In some of these narratives, the science is even propounded as a way to solve a case or to prevent crime.[7] The "Case of Ehlert, the Prussian mate," for instance, includes two extensive, mutually incriminating accounts of the murder of a ship captain by the principle suspects: the mate, Jacob Frederick Ehlert, and the ship's apprentice, Mueller. In each account, one accuses the other of having murdered the captain and enlisting his own reluctant aid in disposing of the body. Although Ehlert was executed, the writer points out that both testimonies "closely tally with the collateral evidence" (65). To solve the case, the writer supplements the insufficient circumstantial evidence and testimonies with new biological evidence in the form of "more numerous measurements" (74). This, he concludes, reveals that Mueller's head houses a greater potential to act on criminal impulses, leading to the conclusion "that both were guilty, Mueller of being the actual murderer, and Ehlert consenting to the deed if not assisting in it." Physiology, the writer finally suggests, offers more reliable information than the relative persuasiveness of personal testimony because, whereas people have motives, physical characteristics do not. Thus, if an investigator uses "phrenological development as a key," it is possible to reliably reconstruct a criminal's past behavior in the absence of physical or testimonial evidence (76).

Much like detective fiction, phrenological case studies aim at cultivating the reader's recognition of a hidden truth that suddenly alters the significance of one's prior knowledge of an individual. The obvious rhetorical aim of this textual practice is to substantiate the validity and practicality of the science, but the foundational assumption of the argument is that there is potentially more to the character of the examined subject than formerly known or suspected. The ultimate object of physiological examination is an entirely new way of understanding human identity as something potentially hidden or obscured from its social context.[8] What the phrenologist purports to do is to solve or explain a crime by examining an entirely different type of evidence: signs of essential identity and criminal potential rather than evidence that something occurred. As one correspondent to the *Phrenological Journal* explains, "offences have a deeper origin;—namely, in organs and excitements, which, through their predominance, produce a certain disposition of mind that impels the individual with extraordinary force to crime" (qtd. in Mittermaier and Combe, 4).[9] The phrenological detective replaces clues and testimony with dispositions and temperament, and motive in these narratives has less to do with the criminal's personal circumstances and more to do with satisfying an internal compulsion.

Phrenologists stressed the ability to detect latent criminal tendencies as one of the most practical applications of the science. They stressed that conventional signs of character, such as attending to reputation, status, and even interpersonal interaction, failed to accurately reveal a person's true nature. As a correspondent to *The Lancet* explains, if the infamous murderer François Courvoisier had applied to be the keeper of a lunatic asylum, "the history of his past life would not have excited suspicion, and testimonials of a high character could have been produced. But what a different tale is told by his cast," which shows "a *secretive* organ sufficiently powerful to prevent any display of the latent disposition . . ." (Hytche 191).[10] In this way, phrenologists attempted to legitimize their science by producing anxiety about traditional modes of assessing character, and then promising to alleviate that very anxiety by offering physiological assessment as the solution. In a sense, these narratives take the form of a mystery, the "fundamental principle" of which "is the investigation and discovery of hidden secrets, the discovery usually leading to some benefit for the character(s) with whom the reader identifies" (Cawelti 42). What phrenological discourse about criminality purports to do is to benefit society by detecting the secrets of identity that are otherwise hidden by self-presentation and personal narrative.

The ethical implications of detecting latent criminality are clearly troubling, and the problematic nature of a biologically based assessment of future potential is addressed at more length in the final chapter of this book. From a narrative perspective, however, simply the idea of latent deviance, completely separable from observable behavior and social context, increases the threatening potential of the sensational villain. If the most salient shift in conventions from gothic to sensation fiction is from the foreign to the familiar—from the Italian castle to the domestic middle-class home—the physiological criminal increases the threat of the familiar even more. The criminal is no longer the starving man on the street or the orphan who has fallen in with thieves; he is now the prosperous, upwardly mobile middle-class man whose only motive to murder, steal, and manipulate comes from an inner compulsion rather than an external need.

It is the potential of this latent, unpredictable threat that Braddon's novel exploits, repeatedly stressing the difference between North's public persona and his latent capabilities. Although introduced with the ironic observation that "he must have been a very good young man, for his goodness was in almost every mouth in Slopperton," by the end of the first chapter, the narrator encourages the reader to question that communal assessment by taking a closer "look" at North: "Look at his face by that one candle; look at the eyes, which are steady now, for he does not dream that any one is watching him—steady and luminous with a subdued fire, which might blaze out some day into a deadly flame" (7, 10). The injunction underscores North's dangerous predispositions, currently "subdued" but nevertheless expressing a "deadly" potential. The repeated imperative to "look" demands the reader enter a position of knowing characterized by close visual observation of the potential criminal's physical features, and more specifically, what these features express in an unguarded moment, before he can assume the social mask that he publicly maintains. By the end of the chapter, the privileged position of "knowing" has been modeled twice: first by the phrenologist, and then by the narrator's injunction to the reader. In both cases, knowledge results from a close scrutiny of North's head. The narrator prepares the reader, early on, to see criminality as something beneath the social surface and beyond the perception of the untrained (or not omniscient) eye. Further, it draws a sharp distinction between this minority position and that of the univocal crowd.

As the novel progresses, the narrator makes use of this oppositional relationship between public power and a more private, disempowered knowing subject to expose the fallacious assumptions about what counts as truth within systems of discipline. According to Foucault, a "'general politics' of truth" includes "the status of those who are charged with saying what counts as true," the "techniques and procedures accorded value in the acquisition of truth" and the "types of discourse which it accepts and makes function as true" (*Knowledge/Power* 131). Within each of these arenas—status, procedural practice, and discourse—Braddon's novel pits accepted representatives of truth against marginalized figures and discounted methodologies. The central conflict of the novel is between the knowing outsider and the "regime of truth" he must overcome by finding ways to make what is already reliably known count as truth in a system that has rendered such knowledge invisible.

The overvaluation of institutional status is one of the key aspects of truth the novel questions. In the portion of the narrative addressing Richard's arrest, trial, and incarceration, Braddon emphasizes the community's assumption that the police are synonymous with correct knowledge. The narrator reports that during Richard's interrogation,

> head officials of the Slopperton police, attired in plain clothes, went in and out . . . though none of them ever spoke, by some strange magic a fresh report got current among the crowd. I think the magical process was this: Some one man, auguring from such and such a significance in their manner, whispered to his nearest neighbour his suggestion of what *might* have been revealed to them within; and this whispered suggestion was repeated from one to another till it grew into a fact, and was still repeated through the crowd, while with every speaker it gathered interest until it grew into a series of imaginary facts. Of one thing the crowd was fully convinced—that was, that those grave men in plain clothes, the Slopperton detectives, knew all, and could tell all, if they only chose to speak. And yet I doubt if there was beneath the stars more than one person who really knew the secret of the dreadful deed. (40)

Braddon draws a distinction between two kinds of knowledge in the passage: knowledge of the actual event possessed only (at this point) by North, and another order of knowledge that the crowd spontaneously generates. The extraordinary phrase "imaginary facts" exposes the potentially fictive basis of factual knowledge while also highlighting how the imagined can assume the aura of the empirical. In addition, the passage demonstrates that what inaugurates and legitimizes the communal construction of "imaginary facts" is the intermittent presence of authority. The town not only conflates knowledge with status but also ascribes to the police the power of omniscience: they "knew all, and could tell all." Braddon thus casts the police detectives as unreliable counterparts to the narrator's own omniscience, and the community as mistrusting and misguided readers of "truth."

Peters's perspective opposes these mistaken but powerful authorities, but the investigation discredits him on the basis of his lack of institutional status. Braddon introduces him as a "dumb [mute] man [who] was a mere scrub, one of the very lowest of the police force, a sort of outsider and *employé* of Mr. Jinks . . ." (29). Based on his "outsider" position, Jinks challenges Peters's assertion of Richard's innocence, saying, "What do *you* know about it, I should like to know? Where did you get your experience? Where did you get your sharp practice? What school have you been formed in, I wonder, that you can come out so positive with your opinion . . . ?" (30) Stressing Peters's lack of credentialing experiences associated with cumulative institutional instruction, Jinks dismisses the conclusion as necessarily baseless, moving the question from the object of knowledge (the evidence of Richard's guilt or innocence) to the relative status of that knowledge's representative.

Despite Jinks's objections, Peters's observational skills have been honed by a lifetime of experience. Because he is mute and communicates in sign, people assume he is deaf and ignore his presence even when discussing sensitive topics. Peters's advantage depends upon a public perception that underestimates his actual abilities, and his social invisibility parallels his relationship to the police force, which also discounts his observations. Although the advantages related to his disability give him the upper hand in the art of detection, his knowledge does not help him build a career or an acceptable case. To do the latter requires an extra step: the translation of accurate knowledge into a form of institutionally recognizable knowledge.

Braddon contrasts Peters's discounted investigative techniques to the procedural practices of the police and prosecution, underscoring that although the detective's methodology is more accurate, it lacks authority. The official investigation builds its case against Richard with circumstantial evidence, primarily the blood on Richard's sleeve and the money in his possession. Knowing the influence such evidence will wield, Peters decides not to oppose the police, later recounting, "I seed that the case was dead agen him . . . only enough to hang him, that's all" (244). In contrast to circumstantial evidence, which depends on making inferences about an individual's actions, Peters uses the direct evidence of physiological "tells." In a chapter largely devoted to explaining his observational technique, Peters explains,

> Now, a cove what's screwed up to face a judge and jury, maybe can face 'em, and never change a line of his mug; but there isn't a cove as lives as can stand that first tap of a detective's hand upon his shoulder. . . . The best of 'em, and the pluckiest of 'em, drops under that. If they keeps the

> colour in their face . . . the perspiration breaks out wet and cold upon their for'eds, and that blows 'em. But this young gent . . . his colour never changed, and he wasn't once knocked over. (243)

Although Peters does not witness the crime itself, he directly observes the physiological signs that betray a suspect's past criminal experience.[11] While keenly aware that a criminal might be able to psychologically prepare himself in certain circumstances, Peters stresses that in the moment of being apprehended the body cannot lie. Facial expressions, sweat, sudden changes in color—these somatic responses cannot be consciously controlled, and thus provide a more accurate index to truth than does circumstantial evidence.

Peters identifies North as the murderer using this technique, while also remaining aware that this certain knowledge has no authority. Peters relates: "Now, what did I see in his face when he looked at me? Why, the very same look that I *missed* in the face of [Richard]. . . . The very same look that I'd seen in a many faces, and never know'd it differ . . . the look of a man as is guilty of what will hang him and thinks that he's found out" (245). This telling instant, which Peters recalls as the key "point in his narrative," mirrors the narrator's earlier injunction that the reader "look" carefully at North's face to find evidence of possible deviance (245, 10). In identifying North as the murderer in this way, Peters assumes the knowing position occupied by the narrator and reader since the first chapter. This physiological tell operates in much the same way as blushing, which Mary Ann O'Farrell claims frequently appears in the nineteenth-century novel to "render body and character legible" by "evad[ing] the constructive capacities of gesture, disguise, and will" (4). Like a blush, the somatic response to guilt prompted by the survival instinct indexes the body's involuntary betrayal, and is therefore more credible than are forms of sociable self-expression. Working from the Foucauldian premise that the confession is one of Western civilization's most privileged forms of truth, O'Farrell suggests that the blush also "performs a somatic act of confession," prompting productive social exchanges (5).[12] As represented in this novel, however, the somatic confession of the criminal is not valued by anyone but Peters, and then only in an oppositional relationship to the techniques of the police and prosecution. If the voluntary confession marks the subject's complicity with and desire to be reconciled to the larger social body, the involuntary confession marks a fissure in the operating procedures of discipline. In fact, it represents that individual's natural desire to escape discipline and, in its very ephemerality, broadcasts the criminal's implicit belief that he can. What Peters registers is a confession that has no value as truth because the technique through which that truth is acquired has not been systematized (and perhaps cannot be systematized) by power. As Peters puts it, "as you can't give looks in as evidence, this wasn't no good in a practical way" (246). This stands to reason: the expression on a face at such a moment is ephemeral and nonreproducible, Peters's direct evidence is one remove from the criminal act itself, and the reliability of the testimony depends on the observer's credibility. Nevertheless, the narrative validates this methodology and uses this moment of physiological recognition to initiate the detective's investigation. Also, Peters's certainty in the physiological observation creates the conditions for suspense in a novel that identifies the murderer from the outset. Rather than relying on the hidden secrets of the criminal's identity or motives, the novel creates tension through dramatic irony. The mystery's solution is not to prove to the reader that North is guilty and Richard is innocent—it is to prove this to the satisfaction of the institutions with power.

On the one hand, Peters's ability to bring North to justice depends entirely on his physiological observations; on the other, this *most* important evidence is the least valuable because it is inadmissible in court. Witnessing North's physiological confession is what compels Peters to gather evidence connected to him, rather than any other individual. In a way quite different from the conventions of later detective fiction, this story begins with the criminal's (involuntary) confession, and then details the accumulation of circumstantial evidence only to legitimize that earlier conclusion (rather than accumulating clues to determine the criminal's identity and provoke a final confession). Peters builds his case by amassing an enormous collection of circumstantial evidence, including an Indian coin that Peters witnessed North toss at a public house, a matching coin found at the scene of the crime, and the testimony of a chemist who once sold North hair dye. Braddon presents the case in court as so much red tape, requiring "the twistings and windings of a very complicated mass of evidence . . ." (393). The evidence is necessary but excessive—an overcompensation for the incompetence of the police and prosecution the last time the case came to trial. "Eight years of that young man's life," Peters observes, "has been sacrificed to the stupidity of them as should have pulled him through!" (359). Given that the solution to the case hangs on a contingency—the coincidental encounter between Peters and North—this statement must refer to the "stupidity" of Jinks in discounting the instinctive assessment of Richard and the prejudice of a system that undervalues or ignores procedural techniques counter to its own.

What underlies and sanctions these institutional failures is a pervasive discourse of character and how it is reliably assessed. If the novel has a didactic point, it is that society attends to the wrong signs when reading character, trusting too firmly in past behavior and reputation. In this respect, North and Richard are diametrically opposed: the town of Slopperton regards the former as a notable and admirable citizen, and the latter as a disgrace. The town bases its assessment of each on their respective life narratives, which for North is one of impressive upward mobility. Through industry and hard work, North moves from orphan "to the Sunday-school teacher; from the Sunday-school teacher to the scrub at Dr. Tappenden's academy; from scrub to usher of the fourth form; and from fourth form usher to first assistant . . . [all] were so many steps in the ladder of fortune which Jabez mounted, as in seven-leagued boots" (8). Watching and approving of his progress and his "meek," "mild," "lamb-like" character, the citizens "praised and applauded this model young man" and "many were the prophecies of the day when the pauper boy should be one of the [town's] greatest men" (9). As Andrew Mangham observes, North represents the popular Victorian cultural ideal of the "self-made man" popularized by such books as Samuel Smiles's bestselling *Self-Help* ("'Murdered at the Breast'" 31).[13]

North, however, is both a self-made man and a critique of this ideological fantasy—particularly insofar as it associated upward mobility with good character. In his chapter on character in *Self-Help*, Smiles argues that a direct correspondence exists between professional success and a good reputation. Using the example of the successful politician Francis Horner, he claims, "It was the force of his character that raised him," since Horner lacked the advantages of rank, wealth, or even especially noteworthy talent. According to Smiles, a bad character breeds distrust, whereas a good character wields a tremendous influence over others. As he puts it, "That character is power, is true in a much higher sense than that knowledge is power" (316). Smiles further claims that a truly good character cannot be convincingly counterfeited, which is crucial to the efficacy of his philosophy, since the ability to simulate a virtuous identity would mean someone could reap the material benefits from a good character while using this identity to mask nefarious intentions.

In North, however, Braddon decouples character from power and social mobility, locating his success entirely in his ability to compellingly assume the appearance of a man worthy of trust. The narrator magnifies the threat of the criminal element by playing on the disjunction between North's public persona and private motivations. In his ascension from pauper orphan to usher, North simulates the values associated with the middle class: piety, humility, and industriousness. He consequently receives a "high character . . . from the Board of the Slopperton Union," which leads to a position at Dr. Tappenden's academy (389). Similarly, when he moves to France and makes inroads into the upper class, he convincingly adopts aristocratic manners and is thus assumed to possess gentlemanly values. For instance, after North blackmails a lady's maid, the narrator writes that she

> watche[d] [his] receding figure with a bewildered stare. Well may [she] be puzzled by this man: he might mystify wiser heads than hers. As he walks with his lounging gait through the winter sunset, many turn to look at his aristocratic figure, fair face, and black hair. If the worst man who looked at him could have seen straight through those clear blue eyes into his soul, would there have been something revealed which might have shocked and revolted even this worst man? Perhaps. Treachery is revolting, surely, to the worst of us. The worst of us might shrink appalled from the contemplation of those hideous secrets which are hidden in the plotting brain. (130)

The passage is similar to the famous description of a beautiful prospect in *Lady Audley's Secret* (1861), which that narrator renders uncanny by relating how "treacherous murders" and "violent deaths" have been committed on rustic country landscapes commonly "associate[d] with—peace" (54). Whereas in *Lady Audley's Secret*, the hallmark of the sensational threat emerges from importing the dangers of an externalized menace into the pastoral, and later the domestic, setting, in *The Trail of the Serpent*, the narrator locates the unsettling threat in North's innate identity. Even the lady's maid who has just experienced his manipulations firsthand cannot reconcile his actions with his appearance and mien. North's repulsive "treachery" does not refer to any betrayal of trust (since they just met) but to his deceptive nature, which hides the intentions of his "plotting brain" from a credulous public. Braddon makes clear that what makes North a threatening figure is the general indecipherability of his nature, rendering him a latent criminal who consolidates power by conforming to recognizable social paradigms.

The same public discourse of character that places North above suspicion condemns Richard, and in this respect, the narrative emphasizes the inverse relationship between the two. For nineteen years North behaves well, but he is innately wicked; during the same period, Richard behaves poorly but is innately good. Before he was charged with murder, Richard had accumulated numerous debts, taken up with bad companions, and deserted his loving mother for seven years (12). The council for the prosecution takes advantage of this history, drawing attention to his past and forcing his mother to admit he had a "reputation of being a scamp—a ne'er-do-well" (57). In a sense, Richard is on trial for his character as much as he is for the murder of his uncle: as the narrator points out, even before the trial "Slopperton had but one voice—a voice loud in execration of the innocent prisoner . . ." (42). Yet, Braddon also draws a distinction between public identity and an essential self that exists entirely separate from personal action. As his mother attempts to explain on the stand, "Apart from his wild conduct, he was a good son. . . . Naturally he had a most excellent disposition . . . dogs followed him instinctively . . ." (58). This is a truly remarkable statement: it suggests that character is separable from behavior, and that repeated actions are not an accurate index of a person's identity. At this point, the only proof of Richard's innocence rests in an appeal to an innate self that might supersede his actions, challenging his reputation as "Daredevil Dick" with an appeal to a more foundational nature (14). Independent of a context that might give these observations the weight of actual evidence, however, such details are abstract and vague—the prosecution mocks the impracticality of ever completing a trial where such "minute" descriptions of "dispositions" might be considered evidence. Nevertheless, the novel posits character as entirely independent of experience and not necessarily reflecting one's personal narrative.

Opposing a discourse of criminal character anchored in experience, Braddon introduces a rival discourse anchored in physiological signs of deviance. Throughout, the narrator supplies physiological commentary on North, and in doing so, models a diagnostic perspective. Observing North in his room after Harding's murder, the narrator comments that he "is as calm-looking as ever" but adds parenthetically, "that absurd professor of phrenology had declared that both the head and face of Jabez bespoke a marvellous power of secretiveness." In describing his pacing of the schoolroom, the narrator similarly observes that "those crotchety physiologists call [his] light step another indication of a secretive disposition" (73). The ironic tone underscores the devalued and rejected status of phrenological and physiological knowledge while also emphasizing how these essential traits offer the only sure clues to North's character. In phrenological discourse, the faculty of "secretiveness" does not specifically refer to keeping a secret but rather to an individual's capacity to repress involuntary reactions that reveal one's true nature. George Combe notes that secretiveness is particularly large in criminals and actors: in the former, it produces "an inward feeling of extreme secrecy, lessens the fear of detection, and thus indirectly favours the commission of crime" (78). In actors, it provides "the power of practising a conscious duplicity, a talent necessarily implied in the representation of a variety of characters . . ." (*Elements* 78–79). North is both a criminal and an actor—or, rather, his ability to act the part of pious usher or stately gentleman puts him in the position of committing a wide variety of crimes without arousing suspicion. As the narrator comments, North possesses "the perfection of the highest art of all—the art of concealing art" (258).

North's essential character allows him to escape all codes of social legibility, but the biological indications of this character escape the system of the social. Paradoxically, the very trait that makes him a master criminal also betrays him. Besides the phrenologist and Peters, the only other character able to penetrate North's social mask is Laurent Blurosset, a "man of science" (147), "a physiologist" (308), and "a chemist" (151), who nevertheless also finds his knowledge devalued by a public that continually questions his methods (148). When North (then going by the alias Raymond Marolles) asks Blurosset for a poison to aid in experiments on rabbits, the scientist secretly substitutes a sleeping potion. Later, he explains that he was able to predict the criminal's intentions by applying his physiological knowledge to a reading of his assumed character. After clarifying that he is a "physiologist [whose] head had grown gray in the pursuit of an inductive science," he explains, "Science would indeed have been a lie, wisdom would indeed have been a chimera, if I could not have read through the low cunning of the superficial showy adventurer, as well as I can read the words written in yonder book through the thin veil of a foreign character" (308). The comparison to translation highlights the necessity for a specialized knowledge—here, physiology—to render North's character legible as what it is, rather than what it appears to be. The play on human character and reading "a foreign character" also aligns Blurosset's scientific perspective with that of the narrator, who has been drawing the reader's attention to the physiological clues that might, if properly valued, reveal North's identity to the world. The physiologist is the only character to directly thwart North's plans, saving a man's life by identifying criminal intent through a reading of essential character and acting on that knowledge.

Blurosset, Peters, and the phrenologist, then, are the only three characters to penetrate North's social mask, and they share two important traits: they all possess physiological knowledge, and they all lack a legitimate position within their respective domains. The townspeople dismiss the phrenologist as "an impostor" (7), the scientific establishment suspects Blurosset of being "a charlatan" (278), and the police see Peters as "a mere scrub" (29). Physiological discourse functions to generate suspense, with these characters operating as an extension of the narrator's omniscience into the world of the story. They "see" in the way the narrator sees, which is in opposition to the prevailing perspectives of the town, the police, the court, and polite society. This arrangement both produces and releases tension in the novel, from the phrenologist's implicit warning to Blurosset's prevention of a future crime.

Ultimately, however, these unconventional techniques that guide the physiological detectives to correct conclusions about human psychology have little power to overturn or reorient the dominant ideology. Peters's work is to translate the knowledge he already possesses into a form of truth that "induces [the] regular effects of power"—here, the incarceration of North (Foucault, *Power/Knowledge* 131). Blurosset acts privately to thwart one of North's plots, but he ultimately assumes the life of a scholarly recluse, pursuing his studies with no concern for professional validation. Braddon writes that "[his] disinterestedness . . . speaks in favour of his truth. . . . He asks no better recompense than the glory of the light he seeks" (278). Yet there is a note of futility in this justification; after all, what good is the pursuit and discovery of a scientific truth that remains only personally significant, and never enters into the exchange of knowledge and power that occurs outside the private laboratory?

Albeit pessimistically, the novel distinguishes between two rival forms of truth: one that is essential and correct, and another that is ideological and thus faulty, prejudiced, and bureaucratic. It has become a critical commonplace to judge a novel's subversiveness by its conclusion, and sensation fiction invites such readings because the genre often addresses issues that unsettle traditional views of gender, class, and social relations only to finally reinstall order through a conventional ending. *The Trail of the Serpent* ends conventionally enough, with North dying in prison, Richard being vindicated, and three marriages being celebrated. However, if potential subversiveness is measured by thematic concerns rather than plot—by what the narrator has revealed to the reader rather than what the social world represented in the story values—Braddon's critique stands. In this, her first work of fiction, an appeal to innate identity functions to expose the mechanisms of power that value status, procedural method, and ideological discourse. The sensational effects of the novel depend on exposing the instability of the knowledge systems upon which Victorian social organization is founded. In response to Dr. Tappenden's remark, "To think that I did not know this man!," the narrator comments, "To think that you did not, Doctor; to think, too, that you do not even now . . . know half this man may have been capable of" (115). The possibility of "not knowing"—and the inability to rely on forms of knowledge with which the narrative audience is familiar—is what the novel exploits. In this respect, the sensational effects provoke an anxiety about institutional power that the novel does not contain or resolve: the serpent's trail remains invisible to all but the disempowered few, and only the rare happy accident of the physiological detective and the innate criminal crossing paths can guarantee safety.

Fittingly, at the end of the novel, only the phrenologist receives public validation for his methodology, since he manages to turn "the whole story into a series of lectures, which were afterwards very popular" (396). This detail mirrors the author's own strategy: positing, through biological evidence, the existence of a nefarious character who the public fails to recognize, and then selling that story for a profit. Much like the phrenologist's lecture series, *The Trail of the Serpent* was also very popular, quickly becoming a bestseller (Wolff 99). Thus, at the outset of her literary career, Braddon found phrenology useful for introducing and enabling a certain kind of sensational plot. When she later embarked on what she considered a more serious literary pursuit in *The Doctor's Wife*, she returned to the science to posit the existence of another type of character society had yet to correctly estimate: a woman with innate abilities that far exceed the paltry opportunities available to her sex.

Shortly after the serialized version of *Trail of the Serpent* appeared, Braddon achieved widespread success with *Lady Audley's Secret*, a book that helped to establish sensation fiction as a distinct genre. After writing three more sensation novels, Braddon made her first bid for artistic recognition in 1864 with *The Doctor's Wife*, an Anglicized version of Gustav Flaubert's *Madame Bovary*. She described the novel to Bulwer as "a kind of turning point in my life, on the issue of which it must depend whether I sink or swim" (qtd. in Wolff, 164–65). Although *The Doctor's Wife* failed to achieve the level of critical recognition she sought, it was Charles Dickens's favorite among her novels and possibly inspired Thomas Hardy's *Return of the Native* and George Eliot's *Middlemarch*.[14] Indeed, much like Eustacia Vye and Dorthea Brook, Isabel Sleaford of *The Doctor's Wife* imagines far greater possibilities for her life than circumstances permit and is quickly disillusioned by an unsatisfying marriage. Like Madame Bovary, she derives her romantic notions from her habit of reading novels. Although her marriage to the respectable but dull George Gilbert saves her from the necessity of being a governess, she rightly finds her domestic life tedious, unstimulating, and mundane. When she meets the similarly literary-minded Roland Lansdell, master of Mordred Priory, she misidentifies him as the romantic hero of her dreams, when in fact he is a common seducer. Unlike her French counterpart, Isabel never commits adultery, naïvely believing that her relationship with Lansdell might indefinitely continue through chaste intellectual discussions. In the end, Gilbert's death by typhoid rescues Isabel from her prosaic domestic life, and Lansdell's murder spares her from his continued pursuit. Improbably, Isabel learns that Lansdell named her his heir, and the novel concludes with praise for her far superior management of the estate.

Perhaps because Braddon based *The Doctor's Wife* on Flaubert's *Madame Bovary* (1856) and because, like it, the novel is very much about reading novels, modern criticism has understandably focused on its literary concerns, be they intertextual or generic, and their effects on the reader.[15] As Lyn Pykett observes, *The Doctor's Wife* is "a self-consciously literary novel, in both its conception and execution" (viii). Indeed, beyond its indebtedness to Flaubert, the novel includes numerous allusions to other works of literature, defines itself in opposition to sensation fiction (while also including a character who is a sensation author), and presents Isabel as a character who conceives of her life in distinctly narrative terms. This foregrounding of its own literariness perhaps accounts for the critical impulse to rigorously contextualize the novel in relation to other works of literature, which has correspondingly served to sever the story from its historical context—a tendency that is far from typical in Braddon scholarship.[16] In a sense, this critical perspective replicates Isabel's own preoccupation with fiction and detachment from the world around her.

Yet, much as in *Trail of the Serpent*, *The Doctor's Wife* makes use of the contemporary discourse of phrenology, and it is this aspect of the novel that reveals a more emphatic critique of woman's unequal position in Victorian society.

In this novel, Braddon narrows her institutional critique to only one system of power: that of male privilege. The narrative dramatizes the injustice of unequal access to the public sphere, both literally (in terms of circulating within and engaging with the world in a meaningful way) and intellectually (through a broader engagement with the world of ideas). Throughout, Braddon represents the domestic as carceral: after marriage, Isabel becomes the object of surveillance by her husband, servants, and the town itself, and she receives approval only when she stays within the home and forces herself to engage in domestic tasks. Where this depiction importantly deviates from the Foucauldian model is that Isabel "earns" punishment for her complicity with the social system because complicity is itself a punishment. Further, the reason the domestic emerges as unjustly carceral is because Braddon repeatedly figures Isabel as naturally *un*suited for her social context. Although our postmodern inheritance might caution that "there is no outside" of the system, Braddon posits just such an "outside" by locating it in a woman's innate mental capacities that, simply by existing, expose the feminine domestic as fundamentally unnatural (Foucault, *Discipline* 301).

Critics have been reluctant to claim this novel as a feminist work, probably because Isabel is not, herself, a particularly agentic or overtly political character. This, however, is precisely the point: Isabel's limited access to the larger world necessarily narrows her perspective and denies her a position from which she might accurately assess her circumstances. The narrator, however, has no such limitations, and the most critical observations on women's subjugation appear in the narrator's commentary. For instance, in one frank aside, she makes the argument that Biblical evidence supports sexual equality because women figure prominently in the Gospels: "Amidst all the arguments to be used against strong-minded claimants for the equality of the sexes, I wonder no one has ever urged the evidence of the Redeemer's treatment of those women whose names are eternally intermingled with His history" (285). At another point, the narrator bitterly underscores the injustice of male privilege, noting that even as little children, Isabel's brothers "looked upon [her] with all that supercilious mixture of pity and contempt with which all boys are apt to regard any fellow-creature who is so weak-minded as to be a girl" (115). Such statements, which critics consistently overlook, reveal that the narrator is clearly and explicitly opposed to cultural assumptions about women's spiritual and intellectual worth. Isabel, however, remains unconscious of the systemic nature of sexism, as she only knows that she is unhappy and cannot find any satisfaction in the roles available to her. The distance between Isabel's experience and the narrator's perspective creates the rhetorical conditions necessary for pathos, a structure that requires distance from the one who suffers and the awareness that the suffering is unmerited. As will be shown, by providing a phrenological perspective on Isabel's innate intellectual capacity, Braddon establishes her mind as equal, if not superior, to the minds of the men around her. This early phrenological revelation foregrounds the unjustifiability of the social system that offers her only a few paltry choices in life, all of which are incommensurate with her formidable mental potential. This same information is, however, purposely withheld from Isabel herself because of the socially perceived uselessness of revealing it. In this way, the narrative dramatizes a woman's experience of mental frustration while also pointing to the underlying conditions that constrain Isabel and impede her mental development.

Although *The Doctor's Wife* certainly is a novel concerned with reading and its effects, it is equally a novel about character, where "character" is understood as both a narrative construct and a representation of human psychology. In her letters to Bulwer-Lytton about the novel, Braddon claimed that she was, for the first time, "going in a little for the subjective" by dedicating herself to the art of "character painting" (qtd. in Wolff, *Sensational* 161). Enfolded into her literary ambitions for this novel was also her desire to realistically portray human subjectivity and psychological complexity. Although much has been said about the novel's inclusion of Sigismund Smith, a sensation novelist who descants on the plotting techniques of his genre, comparatively little has been said about Charles Raymond, the benevolent philosopher-phrenologist who is the only character in the story to correctly assess and understand the psychology of the other characters. Like Smith, who is widely recognized as an alter-ego for Braddon, Raymond is also based on a real person: Charles Bray, the Coventry philosopher-phrenologist, with whom Braddon was friends.[17] Taken together, these figures are counterparts who enact, in their respective preoccupations, the two most basic categories of narrative: plot (Sigismund) and character (Raymond). Yet, unlike Sigismund, who the narrator ironizes to distinguish her own more literary narrative from the sensation genre, Raymond is consistently elevated in the novel as the voice of reason and virtue, and this authority stems from his insights into human nature through phrenology. Serving as a sort of proto-psychologist, Raymond's insights into individual identity parallel Braddon's interest in crafting her own novel of character. By aligning Raymond's correct knowledge of human psychology with a physiological science of mind, Braddon lays claim to a knowledge of human character grounded in scientific authority. Yet, whereas Sigismund's commentary on plotting has reference to a literary and professional context, Raymond's philosophy of human character and its implications can be properly understood only in the context of the Victorian science he represents.

As aforementioned, the character Charles Raymond is based on Charles Bray, a successful ribbon manufacturer, philosopher, and owner of the *Coventry Herald*. His Rosehill home served as a kind of salon for freethinkers of the day, including Robert Owen, Herbert Spencer, George Eliot, and George Henry Lewes (Postlethwaite, *Making It Whole* 66–67, 123). Braddon met Bray in Coventry in 1858 and fondly recalled having "long discussions about phrenology and natural religion, conversations which were like the opening of new worlds to [her] . . ." (qtd. in Carnell, 55). Bray was an enthusiastic devotee of the science, which he subsequently made the basis for his philosophical treatise *The Philosophy of Necessity* (1841) (Bray, *Phases* 21).[18] For him, as for many serious phrenological thinkers, the science demanded a dramatic shift in the arena of mental philosophy. As he explains in his autobiography, "The perfection of our thinking depends upon the more or less perfect development of this instrument of thought; in proportion as that [the brain] is defective will be the world it creates within us. The only way to find this instrument of thought, is the phrenological one of comparing function with development" (*Phases* 34). Like most phrenologists, Bray's perspective was only partially deterministic since he held that the "character of man is the result of the organization he received at birth, and all the various circumstances that have acted upon it since . . ." (*Philosophy* 175). Thus, through self-assessment and education an individual can potentially liberate oneself from the "defective" world the mind creates within. In his *Education of the Feelings* (1872), Bray expresses the need for phrenologically informed early education, which might remedy the insufficient instruction given to young women: "Engaged in the frivolous pursuits of the world, introduced into society . . . dressing, dancing, visiting . . . they are obliged to rely upon an ignorant nurse, to trust to old women's tales, for what ought to have been correct knowledge" (21–22). For Bray, this education is not aimed at improving the quality of women's lives but rather is premised on women's importance as future mothers who will be entrusted with improving the mental condition of their children. Nonetheless, by the conclusion of the volume, Bray's tone becomes more radical and revolutionary as he envisions a future in which education will usher in a classless society: "Let but the principles of our constitution be understood—let but education rest upon a proper basis, and the petty distinctions of a savage age, which form the present scale of society, will be abolished" (194–95).

Charles Raymond resembles Bray both in his achievements and his phrenologically informed worldview. Just as Bray was the philosopher of Coventry, an author of books on social improvement and metaphysics, and was frequently involved in philanthropic activities, his fictional counterpart is "the philosopher of Conventford" (206) who writes "grave books, and publish[es] them for the instruction of mankind" (67–68) while also "carrying out . . . philanthropic schemes for the benefit of his fellow-creatures" (67). Also like Bray, Raymond is a "most faithful disciple of Mr George Combe," and phrenology informs nearly all of his interpersonal interactions (66). More radically than his nonfictional counterpart, however, Raymond sees in Isabel a mental constitution that, with the proper education, might render her a remarkably intelligent and captivating individual in her own right. He muses, "That girl has mental imitation,—the highest and rarest faculty of the human brain,—ideality, and comparison. What could I not make of such a girl as that?" (82). Raymond sees in Isabel a remarkable combination of pronounced intellectual organs that call for an education commensurate with their innate strength.

As explained in the introduction of this book, phrenology's dual emphasis on innate capacity and the effect of environmental influence in unfolding that capacity to its fullest extent provided a rationale for expanding opportunities for women, both in terms of education and occupation. In a similar way, Braddon's use of phrenological discourse in the novel serves to highlight the disparity between women's inherent mental capacity and their limited circumstances, and in doing so mounts a social critique of the educational and occupational opportunities available to women. One of the most salient aspects of phrenology was its claim that innate capacity was often at odds with one's position in life, and this aspect of the science supported the capitalistic ideology of the day. Through the self-knowledge provided by a phrenological reading, a young middle-class man could "discover" that he should be a barrister, and through education and hard work achieve his "proper" place in society. Yet, while such a use of the science accommodated the interests of middle-class men who already had power to safeguard and legitimize, women had far more to gain from phrenology because it revealed to them their own complex mental characteristics and capacities. Whereas the distance between innate capacity and external circumstance for a male artisan was one that could be corrected through individual actions of self-study and self-care, the distance between inherent ability and circumstance for a Victorian woman was immense. Further, closing the gap between character and circumstance for a woman required changes in *society* rather than changes in the self. It is this distance—between inherent mental capacity and external opportunity—that Isabel's phrenological assessment highlights and renders tragic.

Understanding Isabel's intellectual potential by way of phrenological theory reveals a great deal about her character and the origin of her frustration with her life. George Combe associates the three organs Raymond identifies in her cerebral development with exceptional capabilities, particularly in creative endeavors. Imitation "gives the tendency, in speech and conversation, to fit the action to the words" (*Elements* 110). Comparison "gives the power of perceiving resemblances, similitudes, differences, and analogies" and "prompts to reasoning" (138). Ideality, which Raymond elsewhere recognizes as "exaggerated" in Isabel, tends to "elevate and endow with splendid excellence every idea conceived by the mind," and is associated with great writers (106). Taken together, what Raymond recognizes in Isabel is her potential to think, act, and achieve great things. As the novel progresses, Isabel's inchoate abilities continually surface, albeit in cramped forms of expression and abortive attempts at individual achievement. Isabel's organ of "imitativeness" is generally linked in phrenological discourse to the profession of acting, and more specifically by Combe to the renowned English actress Clara Fisher, who began acting at the age of six (*Elements* 83). And, before she marries, Isabel frequently daydreams about becoming an actress, imagining that she could simply go to London, display her powers to the manager, and be offered a part for "fifty pounds a-night" (75). Similarly, her faculty of "ideality," which is typically linked with imaginative poets like Shakespeare, Milton, and Byron, finds expression in her desire to become a poet like Letitia Elizabeth Landon. She actually does begin "a great many verses, and spoil[s] several quires of paper with abortive sonnets" (75). Despite her lack of conviction, other characters recognize her potential to succeed in both professions, Landell once thinking that "[i]f they could get such an Ingénue at the Français, all Paris would be mad about her" (213), and Sigismund urging her to "WRITE A NOVEL!" (229). It is perhaps worth noting that both of the occupations for which Isabel initially yearns were realized by Braddon herself, and that her biographer Robert Lee Wolff claims that she depicted her "own girlhood romanticism in Isabel Sleaford" (162).

As her cerebral development makes clear, it is not lack of ability but lack of opportunity that renders Isabel unable to realize her potential. Throughout the novel, the narrator emphasizes the absence of educational guidance: she is "untaught" (98, 309), "untutored" (112), "half-formed" (161), "half-educated" (185, 292). This leitmotif of deficient education clearly shifts the blame from Isabel to her circumstances. Her romantic fancies and mistaken ideas about the world are, in fact, entirely attributable to the fact that she has been left to her own devices. The narrator elsewhere comments that "[i]f there had been any one to take this lonely girl in hand and organise her education, Heaven only knows what might have been made of her; but there was no friendly finger to point a pathway in the intellectual forest . . ." (29). Clearly, the proper person to perform this service is the phrenologist-philosopher Raymond, but he abandons the idea, reasoning that society would not embrace the kind of woman she could potentially become. Immediately after Raymond muses to himself about Isabel's impressive intellectual capacity, he stops himself "with a sigh." The narrator then explains, "He was thinking that, after all, these bright faculties might not be the best gifts for a woman. It would have been better, perhaps, for Isabel to have possessed the organ of pudding-making and stocking-darning. . . . The kindly phrenologist was thinking that perhaps the highest fate life held for that pale girl . . . was to share the home of a simple-hearted country surgeon, and rear his children" (82). What Raymond recognizes is that the life Isabel is most likely to experience will never allow her to realize her innate potential. Yet, whether or not these mental aptitudes "are the best gifts" for a middle-class Victorian woman, Isabel *has* them; what is more, she lacks the "organ of pudding-making and stocking-darning" not merely because they are not prominent but because they do not exist. Although Raymond tries to convince himself that becoming a wife to Gilbert is her best option, he continually returns to her intellectual faculties and mourns her future: "a sort of picture comes into my mind of what she would be in a great saloon [sic] . . . [with] brilliant men and women clustering round her, to hear her talk . . . and then, when I remember what her life is likely to be, I begin to feel sorry for her, just as if she were some fair young nun, foredoomed to be buried alive by and by" (83). Raymond's assessment of Isabel's intellectual abilities and unsuccessful attempt to reconcile himself to her domestic future is a crucial moment in the narrative: it casts Isabel as an object of pity because she must endure a life for which nature has not suited her. Placing a woman with her cerebral development in a domestic context renders the domestic horrific—the psychological equivalent of being "buried alive." This early passage powerfully marshals the material evidence of a woman's innate mental characteristics to challenge the cultural assumption that the most appropriate role, and highest calling, for the middle-class Victorian woman is to become a domestic helpmate.

In phrenological discourse, the degree to which one experiences happiness is an index of the degree to which external circumstances complement one's innate character. As one phrenologist explains, a person who is not naturally suited for a calling "follows his profession as a painful duty," but individuals who pursue occupations "for which their organization is adapted" will "experience pleasure in those occupations, and soon acquire skill, celebrity, and profit from their exertions" (Watson 64).[19] By codifying the relationship between one's unique identity and specialized labor, the phrenological theory of mental aptitudes, for men at least, offers a biological blueprint for vocational success. It also purports to explain dissatisfaction as a necessary consequence of forcing oneself to engage in activities that do not, and never will, come naturally to the worker. In *Phrenological Conversations* (1833), a phrenologist (most likely James Deville, the leading practical phrenologist in London) relates a number of stories about young men unfit for the life they are intended to pursue, positing phrenology as a safeguard against a life of professional failure and personal misery.[20] In one, a solicitor visits with his twenty-five-year-old son who had been "brought up" to his parent's profession. After making a number of "serious errors" in the family business, the father agreed to visit a phrenologist with the understanding that his recommendations would be "binding." After the examination, Deville reports that "if left to the influence of his organization, and his own inclination, religion would be his occupation"—which, it turns out, was precisely what the son wished (7). In another story, he describes a visit by two mothers, accompanied by their sons, who requested him to identify which young man was best suited for the navy, and which for the church. After examining and identifying each, the phrenologist was told that they were "intended in the reverse," prompting him to recommend that they change the proposed courses of study to suit their sons' characters, which they did. Since that time, one had become an "ornament to the British service" (5), while the other was an "ornament to the church" (6). Such anecdotes, quite common in phrenological literature, marshal evidence in support of the claim that acting in accordance with one's natural dispositions will lead to a greater degree of professional success and personal happiness. Regardless of the amount of environmental influence or familial pressure, phrenologists insisted that innate dispositions remain, giving expression to desires that lead to misery if left unsatisfied by one's occupational experiences.

In a similar way, Braddon frequently highlights Isabel's unhappiness with the domestic life her husband, Gilbert, expects her to enjoy, using the idea of innate character to emphasize the futility of reconciling a woman to circumstances for which nature did not fit her. Isabel's character bucks against the ideological expectations of her sex: she dislikes cooking (156), entertaining (116), and sewing (78). The narrator notes that while there may be women who "take kindly to a simple domestic life, and have a natural genius for pies and puddings, and cutting and contriving," Isabel finds it "a wearisome business" (156). Similarly, when Gilbert wishes her to "be happy according to his ideas of happiness, and not her own" by engaging in "stiff little tea-parties," Braddon notes she "could not make herself happy" by doing so (116). The way in which Braddon invokes domestic activities in relation to Isabel's character emphasizes the dissonance between psychology and circumstance. If some women have a "natural genius" for the domestic, the attendant implication is that for Isabel such activities are *unnatural*, and no amount of self-discipline can induce her to inwardly embrace the life to which she is expected to outwardly conform. At the organic level, she is unsuited to her environment, and by casting that unsuitability in physiological terms, Braddon challenges the naturalizing assumptions that subtend the conflation of sex and feminized labor.

Isabel's indulgence in novels, most often read as a form of escape from her tedious life, is also the extreme expression of her only unrepressed capacities. Even before her marriage, Raymond attributes Isabel's tendency to become absorbed when reading to having the phrenological capacity of "too much Wonder," which causes her to widen "her big eyes when she talks of her favourite books, and [look] up all scared and startled if you speak to her while she's reading" (66–67). His observation tallies closely with Combe's observation that those with pronounced organs of wonder become easily spellbound by works of highly imaginative fiction, often exhibiting "a peculiar look of Wonder, and an unconscious turning up of the exterior angles of the eyelashes, expressive of surprise" (*System* 315). Bray writes at great length about this organ in *The Philosophy of Necessity*, claiming that it is responsible for the mind's ability to believe that sense impressions represent actual objects—in other words, it makes possible the belief that the external world exists at all. For Bray, when the organ is considerably developed, it can lead to taking great pleasure from reading "improbable fictions," which proceeds from "their giving to such tales a reality in their own minds which to others they do not assume" (58). This goes some way toward explaining why Isabel thinks that the romantic stories she reads might be realized in her own life, as she views improbable narratives as perfectly possible, investing them with a degree of belief the average reader does not. Her reading habits, then, are the predictable effect of being placed in circumstances that only allow her to exercise a single facet of her most pronounced mental capabilities. An environment that precludes the development of her intellectual abilities but allows an organ of "sentiment" (wonder) to increase unfettered emerges as potentially dangerous. As Anne-Marie Beller argues, Isabel's tendency to view Roland as a sentimental hero rather than a man with sexual designs demonstrates the "inherent danger" of preserving women's sexual innocence "by presenting the threat of immorality as more likely to result from ignorance" (124). Viewed from a physiological context, however, this moral threat is due not only to withholding certain types of knowledge from women but also to preventing them from experiencing different forms of mental engagement. In other words, unexercised and underused faculties might give less desirable mental tendencies free reign. It is not just the content of women's education but education itself—the habits of mind created and sustained through consistent and varied use—that becomes a salient issue.

Throughout the novel, the subject of women's education, both in terms of instructional guidance and reading, continually resurfaces. Yet the fact that Isabel's unchallenged intellectual capabilities surpass the outer limits of the resources to which she has access points to the insufficiency of the education she receives. The narrator explains early on that she had the "half-and-half education which is popular with the poorer middle classes," but when she leaves school, she "set to work to educate herself by means of the nearest circulating library. She did not feed upon garbage, but settled at once upon the highest blossoms in the flower-garden of fiction," which, for lack of variety, she reads over and over again (27–28). As aforementioned, the subject of Isabel's reading practices has been the central issue for modern critics, who focus on the degree to which she is an "uncritical" reader of novels who fails to distinguish between fictional worlds and reality.[21] This is certainly a key element in the novel, but as the narrator continually points out, Isabel's failure to develop a critical perspective is *itself* a product of limited educational opportunities and a narrow selection of reading material. It is worth noting that Isabel takes the initiative to "educate herself" using the only collection at her disposal and is at least critical enough not to "feed upon garbage." To the extent she is able, in the context of a provincial town while married to a non-reading husband, Isabel continues to display a keenness for whatever knowledge she can access.

Unsurprisingly, then, she eagerly accepts Roland Lansdell's plan for her intellectual improvement. Because he is well educated and well traveled, she sees Lansdell as one who can tell her "of books, and pictures, and foreign cities, and wonderful people, living and dead, of whom she had never heard before . . . Mr Lansdell had the keys, and could open them for her at his will . . ." (184). Lansdell also provides her with a great quantity of more varied reading material, which she "eagerly devoured" (186). If Isabel is an "uncritical" reader, this is largely owing to her limited access to certain kinds of knowledge, which, as the narrator also makes clear, depends on the availability of educational instruction and resources for women. Notably, Isabel almost achieves maturity and wisdom on her own through self-education when Lansdell is removed as a romantic distraction and she has unlimited access to his vast library. The library serves as a version of Woolf's "room of one's own," where Isabel's "mind expanded . . . and the graver thoughts engendered out of grave books pushed away many of her most childish fancies" (235). As the narrator claims, if such self-education under these conditions "could have gone smoothly on for ever, I firmly believe that Isabel Gilbert would have, little by little, developed into a clever and sensible woman" (236). By the end of the novel, Isabel does become this woman, albeit through the "chastening influence of sorrow" from the deaths of Gilbert and Lansdell, which transform her from "a sentimental girl into a good and noble woman—a woman in whom sentiment takes the higher form of universal sympathy" (402, 402–03). Taken together, the two statements imply that Isabel's edification is rather dearly bought, as she achieves through painful personal experience a wisdom more easily accessed through a balanced and expansive education.

Through a series of extraordinary events, by the end of the novel Braddon manages to close the gap between Isabel's latent capabilities and her external circumstances. Her husband dies of a fever, and days later, Roland Lansdell dies from injuries sustained from an attack by Isabel's father. Due to alterations made to his will under the incorrect assumption that Isabel would elope with him to Europe, Lansdell leaves the bulk of his fortune and his estate to her. This ending is remarkable not only because it allows Isabel to survive and prosper, but also—far more radically—it places her in a position that allows her to fully demonstrate her innate capabilities. Under Isabel's influence, the property improves dramatically and flourishes to a degree that far exceeds its management under Lansdell. Although he had once attempted to educate his tenants and renovate the property, his failure to convince them of the feasibility of his schemes led him to quickly abort his plan and adopt a life of self-indulgent leisure. Braddon establishes through the musings of Charles Raymond that Lansdell Priory is completely wasted on Lansdell: "If some rightful heir would turn up . . . and denounce my young friend as a wrongful heir, and turn him out of doors bag and baggage . . . what a blessing it would be . . ." (85). Ultimately Isabel emerges as this "rightful" heir, if not by right of birth, by claims of suitability and effective administration:

> The agricultural labourer . . . had good reason to bless the Doctor's Widow; for model cottages arose in many a pleasant corner of the estate that had once been Roland Lansdell's— pretty Elizabethan cottages, with peaked gables and dormer windows, and wonderful ovens, that would cook a maximum of provision by the aid of a minimum of fuel. Allotment gardens spread themselves here and there on pleasant slopes; and . . . a schoolhouse, a substantial modern building, set in an old-world garden. . . . Even the dreaded innovation of steam-ploughs and threshing-machines brought no discontent to the farmers round Mordred. (403)

Not only does Braddon highlight that Isabel surpasses Lansdell in her management of the Priory, she also notes that Isabel's additions to the estate are aesthetically pleasing, progressive, and innovative. Isabel does not merely maintain the operations established by Lansdell, she creatively improves the practical, technological, and educational resources available to others. Through her management of the estate, Isabel finally exercises the full potential of her pronounced faculty of "ideality," which is also linked in phrenological discourse to realizing idealistic aims in the actual world. As Combe remarks, "It is particularly valuable to man as a progressive being. It inspires him with a ceaseless love of improvement, and prompts him to form and realize splendid conceptions" (*Elements* 106–07). As it turns out, Isabel was always more qualified to manage an estate, gifted with innate capacities superior to Lansdell's comparatively unremarkable head, with a brow that "was neither high nor low" and a "cranium" in which both "veneration and conscientiousness were deficient" (128). As Raymond puts it, Lansdell is "brilliant without being a genius. In short, he's just the sort of man to dawdle away the brightest years of his life . . ." (133). While Lansdell's cranium betrays a lack of genius, Isabel's head allows her to be classed among "those gifted beings" who carry the "flaring torches men call genius" (83). Although Braddon distributes these references throughout the text, there is certainly a submerged comparison in the novel between the man who inherits the estate and the woman who deserves to run it. In stark opposition to the fallacious assumption her half-brothers hold, pitying her for being "so weak-minded as to be a girl" (115), Isabel's mind is predicted to be—and by conclusion of the novel, proved to be—more capable of achievement than the mind of a man.

Kate Flint poses the provocative question, "What are the implications of the fact that two men have to die in *The Doctor's Wife*" in order for Isabel to be "rescued by the narrator for an independent woman's life?" (292).[22] One possible answer is that in its improbability, the conclusion is precisely what it seems to be: a *deus ex machina* that alters Isabel's external circumstances radically enough to allow her inherent capabilities to surface in a productive way. Extraordinary events must transpire for only one woman to realize her potential, and only through this rare turn of events can Isabel meet a happier fate than other characters whose innate nature is hampered by social realities. For instance, Gwendoline Pomphrey, Lansdell's cousin, has an innate nature that "would have well become a young reformer, enthusiastic and untiring in a noble cause."

The narrator laments that in such cases as hers, a "bright ambitious young creature, with the soul of a Pitt, sits at home and works sham roses in Berlin wool; while her booby brother is thrust out into the world to fight the mighty battle" (296). The narrator delivers this bitter observation after Gwendoline contemplates with a "yearning gaze" in her eyes, how she would outperform Lansdell in his place: "If *I* were a man," she muses poignantly, "I would be up and doing among my compeers" (296). Unfortunately, there is no *deux ex machina* that might transport Gwendoline directly into the political sphere. For Isabel, however, it was at least within the bounds of reality, if not probability, to close the gap between Isabel's innate nature and external circumstance. Rather than exercising her faculties within the narrow sphere of imaginary worlds, Isabel emerges as one who authors the ideal through her progressive management of an estate. By the end of the novel, Isabel's sphere has literally expanded from the cramped confines of the domestic to the large circumference of Lansdell Priory.

What Isabel Sleaford and Jabez North have in common is that both are underestimated by their respective communities, and even by those who are closest to them. Both possess latent abilities that emerge in ways that overturn expectations grounded in knowledge of their past experiences, and in defying these expectations their characters destabilize the assumptions that underwrite interpretive paradigms about what constitutes individual identity. Braddon's choice to have phrenologists reveal these innate abilities points to her sustained interest in a physiological basis of mind being used to contradict ideological constructs. For North, latent ability pitted against social expectations plays to his advantage as he, self-aware and completely agreeing with his phrenological reading, pretends to be what he is not: a pious youth, a self-made man, a trustworthy banker. As a woman, however, Isabel cannot benefit from her inherent abilities, and try as she might, she cannot force herself to be happy in the limited options available to her sex.

In *The Doctor's Wife*, the destabilization of social categories in relation to gender is more nuanced but also less ambiguous in its implied critique. The novel is not an overtly political text, but it contains poignant moments expressing earnest bitterness about how social conventions prevent women from achieving individual identities. Isabel's constant feeling that there is something more to life than the blank and unsatisfying existence she daily experiences is not unlike the rudderless yearning of housewives described in Betty Friedan's *Feminine*

Mystique (1963). Like those women, who experienced a "strange feeling of desperation" that they cannot explain, Isabel knows what she feels but not why (Friedan 64). The narrative stops short of giving us any sense of how self-aware she becomes about her past experiences, or the nature of the "wisdom" she achieves, instead leaving us with the effects she leaves on the world. Isabel is not herself a radical proponent of the equality of the sexes, but as Friedan remarked of the 1950s housewife, "How can any woman see the whole truth within the bounds of her own life? How can she believe that voice inside herself, when it denies the conventional, accepted truths by which she has been living?" (77). The hand of the narrator lifts her out of her circumstances and aligns her with an occupation commensurate with her capabilities, even if she is, in Kate Flint's words, "unworthy in many ways, to achieve a far happier fate than that of Emma Bovary" (292). By foregrounding her natural capacities through physiological science, Braddon makes a subtle but provocative case that Isabel was always "worthy" of more than what middle-class domesticity offered her: capable of building schools rather than making pudding, of beautifying land rather than a parlor, worthy of innovating agricultural practice rather than darning socks.

Robert Lee Wolff has commented that Braddon's "incisive but unobtrusive social commentary . . . usually escaped the notice of her contemporaries" ("Devoted Disciple" 5). He meant this in terms of the more literary aspects of her sensation fiction, but there is also a sense in which the subversiveness of her fiction flies under the critical radar. There are moments here and there where a character breaks from the central issues at hand to make a startlingly frank comment about women being barred from much professional life. For instance, Robert Audley's assertion that women are "the stronger sex, the noisier, the more persevering, the most self-assertive sex. They want freedom of opinion, variety of occupation, do they? Let them have it. Let them be lawyers, doctors, preachers, teachers, soldiers, legislators—anything they like—but let them be quiet—if they can" (207). Or Valentine Hawkhurst's more sympathetic plea, "O, let us have women doctors, women lawyers, women parsons, women stone breakers—anything rather than these dependent creatures who sit in other people's houses, working prie-dieu chairs, and pining for freedom" (2: 285). Such observations, while remarkably progressive from the mouths of male characters, are nonetheless tangential to their respective plots. *The Doctor's Wife*, however, offers one of the most sustained and straightforward meditations in Braddon's fiction on the circumscribed position of middle-class women in Victorian society. It is to Braddon's credit that she grants Isabel a happier end, who feared as a girl that "perhaps her life was to be only a commonplace kind of existence . . . a blank flat level" when in her heart, "She was so eager to be *something*" (73). Nature attested to the fact that she was capable of being something more than commonplace, and in this novel at least, nature is right.

Chapter 5

George Eliot and Biological Destiny

Unlike the other women writers discussed in this book, George Eliot moved from an initial interest in phrenology to a staunch critique of its social implications. What made the science so inviting to other women—its accessibility and comparative openness to their sex—was less remarkable to her, obviated by personal and professional connections that gave her access to research and discoveries at the forefront of biology and physiological psychology. She also had intimate relationships with Herbert Spencer and G.H. Lewes, two pioneers of evolutionary psychology whose work laid the foundation for a materialist approach to mind in mainstream science. During Eliot's time as assistant editor of the *Westminster Review*, she shepherded important work into print on a variety of scientific subjects, including Spencer's own theory of evolution. Eliot also assisted Lewes in collecting and cataloging specimens for his *Blackwood's Magazine* series "Sea-Side Studies," during which time she developed an interest in the structure and biological processes of seaweed (Hughes 173). Perhaps most impressive in terms of its rigor, after Lewes's death she completed the final volume of his monumental *Problems of Life and Mind*, a work that posited consciousness as the effect of interactions between neural and environmental forces.

Long before her relationships with Spencer and Lewes, however, Eliot had a mentor in the phrenologist-philosopher Charles Bray, whose Rosehill home served as an intellectual sanctuary for her before her professional life began.[1] During the period of her closest intimacy with the Brays (1841–1849), Eliot personally experienced at least three phrenological readings, received instruction from the London phrenologist Cornelius Donovan in 1844, and had a cast made of her skull.[2] Her letters during this time also reflect her familiarity and comfortability with phrenological language: "Mlle. Elise . . . is a German with a moral region that would rejoice Mr. Bray's eyes" (*Letters* 1: 293); "this bump of cautiousness with which nature has furnished me is of very little use to me" (1: 255); "Mr. Donovan's wizard hand would detect a slight corrugation of the skin on my organs 5 and 6: they are so totally without exercise" (1: 193). In the summer of 1851, the Brays introduced her to George Combe, and the two immediately formed a close friendship initially characterized by mutual respect and admiration. Combe was particularly impressed with Eliot's skull, recording in his journal, "She has a very large brain, the anterior lobe is remarkable for length, breadth, and height, the coronal region is large, the front rather predominating . . . she appeared to me the ablest woman whom I have seen . . ." (qtd. in *Letters*, 8: 27–28). Similarly impressed, Eliot effuses in a letter to Sara Hennell that Combe was "an apostle. An apostle, it is true, with a back and front drawingroom, but still earnest, convinced, consistent, having fought a good fight and now peacefully enjoying the retrospect of it" (2: 61).

Eliot's interest in phrenology, however, waned as she grew closer to Lewes, one of the most outspoken critics of the science. In his 1857 article "Phrenology in France" for *Blackwood's Magazine*, Lewes argued that phrenology failed to account for the "unequivocal facts" that contradicted its claims, concluding, "We think its Psychology excessively imperfect. We think its Physiology crude and inaccurate, and far below the present state of knowledge. We think its pretence of reading character a mischievous and misleading error" (673).[3] In the same year, Lewes added a chapter on phrenology to the revised edition of his *Biographical History of Philosophy*. While he acknowledged Gall's contribution to the history of psychology by introducing a physiological approach to the study of mind, he criticized phrenology for its reliance on observational correlation rather than rigorous anatomical research. Bray took exception to this argument, publishing a rebuttal in the second edition of his *Philosophy of Necessity* in which he speculated that Lewes must have based his opinions on encounters with amateurs and charlatans rather than "recognized leaders in the science" (148). In a letter defending Lewes's position to Bray, Eliot conciliatorily states, "I suppose Phrenology is an open question, on which everybody has a right to speak his mind," but adds, "I cannot except you from my general experience of phrenologists.

With regard to their system, they seem to me to be animated by the same sort of spirit as that of religious dogmatists . . ." (2: 403–4).

Although Eliot soured on phrenology in the decade following her time at Rosehill, she nonetheless remained interested in one of the science's central preoccupations: the degree to which the body serves as an index of intellectual potential, thereby offering a form of biological future knowledge. In both her short story "The Lifted Veil" (1859) and her 1865 poem "A Minor Prophet," she critiques phrenology for reasons related to the science's claim to predict the mental capabilities of individuals and social groups. Superadded to a belief in the efficacy of phrenological prediction, the central characters of both works also claim to divine the future in more specific ways. Latimer, the narrator of "The Lifted Veil," actually sees his own future experiences through flashes of "prevision," whereas Elias Baptist Butterworth, the "Minor Prophet" of the poem, envisions a future he hopes to manifest through elaborate utopian schemes. In both, Eliot uses phrenology to underscore the delimiting effects of these different forms of predictive vision. In Latimer's case, a phrenological reading correctly gauges his innate intellectual capabilities, but his father uses this information to justify an educational regimen intended to counteract and modify his dispositions. Butterworth, emboldened by his own impressive frontal lobe, sees himself as both prefiguring and preparing the way for a biologically perfected future society that excludes inferior types. In both works, Eliot demonstrates how attempts to forecast and control the future through reductive scientific assessments of the body ultimately tend to create the conditions for a limited and limiting future.[4] By foregrounding the problematic consequences of conflating psychological interiority with physical signs, Eliot criticized the way biological determinism was then being deployed, both publicly and privately, to reorganize and reform society in troubling ways.

While she remained critical of phrenology, Eliot returned to the intersection of the body and future knowledge in *Daniel Deronda*, and she did so in a way that both references and reframes her earlier explorations of these subjects. Unlike Latimer and Butterworth, who posit nature as a determining force, the visionary character of Mordecai actively claims his ability to contour physiological interpretation and rhetorically craft, through it, a future beyond socially inscribed possibilities. In doing so, he engages in a form of strategic essentialism that counteracts the pathologizing effects of discriminatory practices rooted in visual assessments of racial identity. Understanding Eliot's concerns with the ways in which scientific discourse tends to connect physical characteristics to future potential helps to clarify her alternative vision of biological destiny in relation to race in a novel that significantly addresses the role of prophesy and projection in the service of human progress.

In repurposing essentialism for progressive ends, her underlying approach mirrors that of the other women writers in this book, and *Deronda* was not the first time Eliot adopted such a strategy. As shown in chapter 1, feminist phrenologists used the idea of the skull's predictive power to justify enlarging women's freedoms by pointing to their cerebral qualifications for educational opportunities and professional occupations then barred to them. Eliot makes a similar argument in her *Westminster* article on "Woman in France," in which she claims "a physiological basis for the intellectual effectiveness of French women" ("Woman" 56). She argues that the comparatively greater cultural achievements of French women over those in other European countries is due to innate mental characteristics enhanced, over time, by external circumstances conducive to intellectual development. French women possess "the small brain and vivacious temperament which permit the fragile system of woman to sustain the superlative activity requisite for intellectual creativeness," whereas the "larger brain and slower temperament of the English . . . requires a larger sum of conditions to produce a perfect specimen" (55). She locates the "condition" that benefits French women in the open intellectual atmosphere of the salon, where both sexes met on equal footing to discuss literature, science, and art (57–59). Although her reference to large-brained Englishwomen reveals a less radical gambit than that of the feminist phrenologists, she still anchors her approach in innate physiological capabilities that deserve a more congenial medium in which to develop. In fact, she argues that English women be afforded *more* educational and social freedoms than those enjoyed by women in France if they are to compete intellectually. She concludes by evoking the pathos inherent in the disjunction between English women's innate mental ability and impoverished opportunities, stating, "we sympathize with the yearning activity of faculties which, deprived of their proper material, waste themselves in weaving fabrics out of cobwebs," a situation she suggests will be remedied when "the whole field of reality be laid open to woman as well as to man . . ." (80–81). This demonstrates Eliot's recognition of how biological essentialism could be used for reformist ends, albeit in a way that reflects her general perspective on social progress as a slow process of incremental change through time.

Nonetheless, she did see the biological body—and especially its cultural legibility—as importantly intertwined with the way in which society imagines, projects, and manifests its future.

The tendency to link the body's materiality to predictive assessments was particularly central to phrenology's practical application in education. At the end of *The Constitution of Man*, George Combe predicts that in the future the state will require all children to undergo a phrenological assessment and subsequently receive an education designed to make the most of their innate abilities (330). Combe was putting this plan into practice in his personal life in ways that would have come under Eliot's notice during her stay at his Edinburgh home from the seventh to the twentieth of October in 1852. During this period, he records in his journal two important callers who sought his advice on education.[5] One was Dr. Ernst Becker, an expert enlisted by Prince Albert (along with Combe) to improve the mental health of the Prince of Wales.[6] On October 12, Becker visited to report that their educational strategy had resulted in the prince being "greatly improved in his Physiological condition."[7] Two days later, Combe conducted a phrenological examination of the nine-year-old son of the former Prime Minister Lord John Russell, for whom he prescribed a regimen of vigorous exercise to increase mental vitality.

Russell's interest in phrenology's usefulness in molding the future citizenry through education stretched back to the 1830s, when he campaigned for a national system of secular education. Two phrenologists similarly devoted to educational reform shaped his perspective: the first was Thomas Wyse, an Irish M.P. who helped to establish a Board of Commissioners in his home country that became the model for Russell's Committee of the Privy Council on Education (Murphy 15). The second was James Simpson, with whom Eliot was also personally acquainted through the Brays.[8] Simpson was a prolific and well-known author on the subject of education, and he based his recommendations on phrenological and physiological knowledge. In his influential *Philosophy of Education*, he argued that the future of England depended upon the implementation of a national system that could counteract and ameliorate the unchecked immoral propensities and constitutional weakness of the working classes. He writes:

> The manual labourer, whom filth, foul air, muscular and nervous relaxation aggravated by ardent spirits, have combined to predispose to and affect with disease, has had no lesson ever taught him that his weakened frame, predispositions, and

actually formed diseases, will be the wretched inheritance of his children, if he shall become a father. [. . .] He himself derived a tainted constitution, perhaps, from his progenitor, and, with his own actual deteriorations superadded, conveys it to his offspring. A few such generations must extinguish the stock, the very source of such a population. (13–14)

For Simpson, the best evidence for the necessity of a national system could be found in the rapidly degenerating physiological constitution of the greater part of Britain's populace. Simpson's proposal was fairly progressive, advocating free education for all classes and both sexes from the ages of two to fourteen (163, 132). This standardized plan, as Simpson admits, might seem to run counter to the phrenological doctrine that each human possesses a unique combination of innate mental characteristics that qualifies them for a specific kind of training. He clarifies, however, that children should exercise all faculties equally until puberty because a specialized education based on endowment would render a child "a cretin with one faculty working like a blind instinct." Further, he claims a general education serves to bring a particular talent "into bold relief" such that "the start to farther attainments in the marked line of the special talent, will be immediate, well directed, and energetic" (126). Rather than a tailored education, Simpson's approach emphasized bodily exercise, hygiene (95–97), and lessons on subjects ranging from arithmetic to political economy (with a special focus on scientific subjects) (113–23). This curriculum strongly resembles that piloted by Combe in 1830 at his Edinburgh Infant School (Tomlinson 172–77), which Simpson approvingly cites throughout the *Philosophy*. He also refers to Combe's school in evidence given to a select committee (first organized by Russell) on the state of education in 1835 (190–95).

Combe, Simpson, and Russell were thus all major players in educational reform, and they shared a similar vision of popular elementary education based on phrenological principles. Key to the efficacy of a national system was uniformity, so the idea of specialized intellectual training was inimical to their aims. Still, pursuing a focused professional education after reaching maturity was a common theme in phrenological texts, and phrenologists frequently advertised their ability to help young men discover what career was best suited to their natural talents (Figure 5.1).

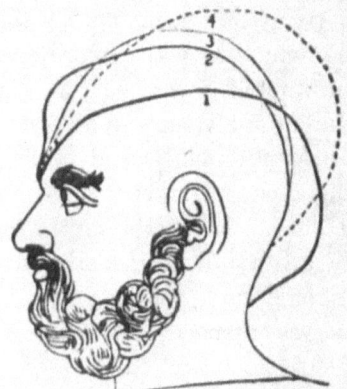

Figure 5.1. Advertisement for phrenological examinations at Lorenzo Niles Fowler's Phrenological Rooms at Ludgate Circus, London. Published in *The Phrenological Magazine* vol. 1, no. 3, 1885.

Unlike elementary education, professional education on the phrenological plan sought to make the most of one's physiological predispositions by placing a youth in a context that encouraged exercising the most pronounced faculties. As the logic ran, if a young man were placed in the environment for which he was suited, he would find work pleasurable and distinguish himself in that field as a matter of course. The inverse of this was training individuals to a career for which they were not naturally suited, an act phrenological advocates viewed as inhumane. Spurzheim belabors this point in his *View of the Elementary Principles of Education*, explaining, "It is a lamentable truth, that few persons stand in the situations for which nature particularly fitted them. This soldier ought to have been a clergyman; that clergyman a soldier; and here we see a shoemaker who was intended for a poet . . . precepts and rules neither bring forth talents nor moral conduct . . ." (206–7). The point of a phrenological reading was to determine one's most marked intellectual characteristics so they might be encouraged and accommodated, ultimately leading to professional success and personal happiness. The version of physiologically based education Eliot dramatizes in "The Lifted Veil," however, runs entirely counter to the pedagogical and professional philosophy supported by phrenological discourse.

In "The Lifted Veil," Latimer narrates the story of his life, up to and including the moment of his death, which he has foreseen. In addition to prevision, Latimer can also hear the thoughts of others, an ability that reveals the petty motivations and shallow concerns of everyone he encounters, eventually forcing him to become a recluse. His knowledge of the future similarly annihilates desire and renders him passive in the wake of his inevitable future. The story is uncharacteristic of Eliot's corpus, both for its use of a first-person narrator and its exploration of extrasensory abilities. Further, Latimer's unmediated access to others causes him to call into question the efficacy of fellowship and sympathy, two of Eliot's most cherished values. Eliot herself recognized the tale as a departure, calling it a "slight story of an outré kind—not a *jeu d'esprit*, but a *jeu de melancolie*" (*Letters* 3: 41). Despite her ambivalence, "The Lifted Veil" has since generated considerable critical interest due to its sustained engagement with a range of contemporary scientific topics, including sensory physiology, microscopy, mesmerism, and telepathy.[9] Very little, however, has been said about Eliot's representation of phrenology in the story, or its contextual significance as the only science in the story that had immediate implications for social policy.[10]

Given this larger context, Latimer's decision to begin the narrative of his past with a phrenological reading dramatizes the possible psychological consequences of a wrongheaded attempt to predict and alter an individual's future. Latimer emphasizes the importance of the encounter in his life by describing his experience with the phrenologist, Mr. Letherall, as calling forth his "first hatred" (6). Latimer's father enlists Letherall's services to ratify a predetermined plan to provide his second son with a "scientific education," a program that suits his own interest in "mining speculations" (6). Infelicitously, Latimer's cerebral development reflects a poetic sensibility completely at odds with the professional future his father desires for him. Rather than encouraging the father to support his son's natural talents, Letherall recommends that he suppress his son's gifts and train up his deficiencies: "The contemplation [of my skull] appeared to displease him, for he frowned sternly, and said to my father, drawing his thumbs across my eyebrows—'The deficiency is there, sir—there; and here,' he added, touching the upper sides of my head, 'here is the excess. That must be brought out, sir, and this must be laid to sleep'" (6).

The line of organs above the eyebrow (number, order, color, weight, size, and form) correspond to spacial perception, mechanical reasoning, and mathematical ability—precisely those faculties put to use in a scientific career. The organs in "excess" on the upper sides of Latimer's skull—presumably wonder and ideality—are typically linked with poetic ability in phrenological texts, with wonder inspiring the imagination and ideality producing a "love of the beautiful" that is "essential to the poet" (Combe, *Elements* 106, 107). In his *Elements of Phrenology*, Combe includes an illustration of the poet Tasso as a prime example of someone possessing both faculties in abundance.[11] Following Letherall's advice, Latimer's father provides his son with an education in the natural and applied sciences in an effort to correct the "defects of [his] organisation" (6). Latimer points out the absurdity of this plan, observing, "I was very stupid about machines, so I was to be greatly occupied with them; I had no memory for classification, so it was particularly necessary that I should study systematic zoology and botany; I was hungry for human deeds and human emotions, so I was to be plentifully crammed with the mechanical powers, the elementary bodies, and the phenomena of electricity and magnetism" (6). This program of systematically stultifying innate talents and forcing the development of impoverished faculties agrees in no way with the educational ideas of Combe, Spurzheim, or Simpson, nor with the general principles of phrenology that held that while individual character can be somewhat modified through exercise, character essentially remains the same. From a phrenological perspective, Letherall's advice is unsound, perpetuating what Spurzheim calls the "great[est] harm to society," that of "placing individuals in professions and situations for which they [are] unfit, not only through the want of some necessary faculties, but also through the inordinate activity of some of the opposite ones" (199). Either Eliot was ignorant of the dominant phrenological discourse on education, or she purposely styled Letherall as a bad phrenologist—either way, the prescriptive education Latimer receives negates the assumption that definitive knowledge of one's inherent capabilities is always liberating and useful.

Two other elements about this scene are curious given Eliot's disenchantment with the science. First, the information Letherall offers is unnecessary given that the father had already settled on the kind of education his son would receive before consulting the phrenologist. Equally strange is the correctness of the reading: after describing how Letherall identified him as a "sensitive boy" with an "excess" (6) in the poetic region of his brain, Latimer confirms that his "nature was of the sensitive, unpractical order," and that he did possess a "poet's sensibility" (7). This raises the question of why Eliot would include a scene both superfluous to the plot and that endorsed the validity of a science she found pernicious and unscientific. Given the story's central thematic concern, however, the phrenological opening makes perfect sense: as its title indicates, the story explores the effect of knowledge typically barred to human consciousness, and as the epigraph makes clear, this barrier is fortuitous: "Give me no light, great Heaven, but such as turns / To energy of human fellowship; / No powers beyond the growing heritage / That makes completer manhood."[12] And, indeed, the "light" Latimer receives from clairvoyance and telepathy renders him despondent, miserable, and disconnected from others. The phrenological encounter that precedes his extrasensory abilities is thus the first of three veils that, for the sake of personal happiness and "human fellowship," should have remained unlifted. Unlike prevision and telepathy, however, only phrenology was then being used to restructure and reform English institutions.

The scene is also significant because it foregrounds a key tension between two determining forces—nature and nurture—that importantly operate at cross purposes throughout the story. Ultimately, knowledge of his innate identity reconciles Latimer to a life of thwarted desire in the wake of the counterbalancing pull of internal drive and external experience. He depicts the private education he receives as a form of psychological abuse precisely because it works against his formally acknowledged natural predilections and abilities. To produce the desired results, he is kept "under careful surveillance" and must resort to reading "Plutarch, and Shakspere [sic], and Don Quixote by the sly . . ." (8, 7). Despite his tailored curriculum, Latimer does not improve in either scientific or humanistic thinking, as he lacks the ability to understand his studies and the necessary training to become a poet. Combating nature with nurture does not remove his innate capacities but rather deforms them. "My nature," he explains, "grew up in an uncongenial medium, which could never foster it into happy, healthy development." Underscoring the physiological basis of his character, Latimer likens his "nature" to an organism placed in an inhospitable environment, rendering his mature character the maimed product of a failed experiment. By hampering his nascent intellectual capabilities, his education leaves behind compulsions that his stunted faculties cannot properly express. As he puts it, he has a "poet's sensibility without his voice—the poet's sensibility that finds no vent but in silent tears . . ." (7). In a passage Eliot deleted from the 1878 Cabinet Edition but that appeared in the original *Blackwood's* version, Latimer makes his conviction of educational mistreatment more explicit, remarking, "I have been led to conclude that the only universal rule with regard to education is, that no rule should be held universal, a good education being that which adapts itself to individual wants and faculties" (26). Perhaps removed because it is so uncritically phrenological, the sentence nevertheless underscores the potentially damaging psychological effects of infelicitous environmental conditioning.

Ultimately, the phrenological diagnosis and counteractive schooling it legitimizes operate as Latimer's first great lesson in power, convincing him that he will always be at the mercy of external forces. Latimer dramatizes the pathos of this disproportional dynamic in his account of the interaction, explaining,

> Mr. Letherall was a large man in spectacles, who one day took my small head between his large hands, and pressed it here and there in an exploratory, suspicious manner—then placed each of his great thumbs on my temples . . . and stared at me with glittering spectacles. . . . I was in a state of tremor,

> partly at the vague idea that I was the object of reprobation, partly in the agitation of my first hatred—hatred of this big, spectacled man, who pulled my head about as if he wanted to buy and cheapen it. (6)

The density of the references to Letherall's size in contrast to Latimer's "small head" emphasizes the degree to which he felt physically overmastered while being disparagingly appraised. The encounter also operates as a compressed version of the larger structure of control the father exerts over his son: in the same way Letherall treats Latimer's head "as if he wanted to buy and cheapen it," the father treats his son's future as a potentially lucrative investment that he ultimately debases through environmental influence. The repeated mention of "spectacles" intensifies the visual aspect of the experience, but it also recalls the father's interest in "speculation," an economic enterprise based on profiting from a change in market value as opposed to typical forms of investment or trade. In a similar way, the father hopes to enrich himself by changing the market value of his son—a plan that proves to be as risky as most speculation schemes, although Latimer pays most dearly.

Latimer's subsequent and repeated denials of personal agency rehearse this formative lesson in the determining power of a deformed nature. When he is finally free to pursue poetry as an adult, he claims his "half-womanish, half-ghostly beauty" was entirely "too feeble for the sublime resistance of poetic production." He similarly recalls, "I saw in my face . . . nothing but the stamp of a morbid organisation, framed for passive suffering" (14). In fact, the only time Latimer briefly believes he might actually be able to become a poet is after a severe illness, which he speculates might have "wrought some happy change in [his] organisation—given a firmer tension to [his] nerves" (10). For Latimer, only a physiological transformation that literally changes the structure of his brain could possibly produce a change in fortune. When he subsequently discovers that the illness gave him clairvoyant abilities rather than the poetic talent for which he hoped, he similarly speaks in biological terms, speculating that his prevision might be "a disease . . . concentrating [his] brain into moments of unhealthy activity" (12), and characterizing his telepathy as a "diseased consciousness" (14). Externally and internally, Latimer views himself as incapable of escaping the somatic limitations that have already determined his fate by dictating a future of fixed limits.

Latimer's definitive knowledge of his own character demonstrates how unaccountable an individual's response may be to a plan that sacrifices unnumbered possibilities for a modicum of preemptive control.

Whereas the adverse effects of biological foreknowledge center only on one aberrant individual in "The Lifted Veil," Eliot is more explicitly critical of phrenology when it is invoked as a tool for contouring entire populations based on predictive assessments of mental tendencies. Combe advanced a shockingly determinist plan for preventing crime in this way in his article "Criminal Legislation and Prison Discipline," which was submitted to the *Westminster Review* during Eliot's editorship. In it, Combe earnestly recommends state-enforced phrenological examinations at birth in order to determine the likelihood of each British citizen to commit a violent crime in the future. After explaining that the anterior lobe is the seat of the intellectual and moral faculties, whereas the posterior lobe houses the animal propensities that lead to deviant behavior, Combe categorizes British society into three groups: normal individuals, with proportionally larger anterior lobes; corrigible criminals, with small anterior lobes; and incorrigible criminals, with extremely small anterior lobes (430). For those belonging to the incorrigible class, he calls for lifelong imprisonment, and for corrigible criminals he recommends an individualized regimen of training and instruction calculated to normalize the subject by exercising, and thus increasing, the size and power of the anterior lobe (431). Combe underscores that the particular value of his plan is its ability to locate, incarcerate, and rehabilitate criminals before crimes have been committed, explaining that an "effect of persisting in disregarding the influence of the organism [the brain] is, that though in many cases the coming event of violent injury casts its shadow before, this premonition is unheeded . . ." (427). In his plan, Combe advances the twin overarching aims of phrenological social reform through predictive measures: fix the individual whose nature can be altered, and remove those who are beyond saving from the population. By referencing the "physical constitution" (i.e., the skull) as the quantifiable, unambiguous indication of deviancy, biological determinism becomes the mechanism for reworking society's fabric (424). In this way, Combe promoted phrenology as a kind of scientific second sight, which reveals a potential future that can be passively accepted or actively erased.

Unsurprisingly, Eliot was reluctant to accept the article—a resistance Combe had clearly come to expect. In a letter to Eliot in 1852,

Combe alleged that the *Westminster* was purposely ignoring the subjects of phrenology and mesmerism. Eliot replied that the accusation was false, but added, "I think you will agree with me that the great majority of 'investigators' of mesmerism are anything but 'scientific.'" This rebuttal follows Eliot's claim that the *Westminster* sought only "the best thought and the best writing on the most important topics" (8: 41). Preemptively hostile, when he submitted the article on criminal legislation the following year, Combe commented, "You and Mr. Chapman may feel quite at ease about rejecting it, if found too phrenological and technical, for I shall print it as a pamphlet and distribute it, if you reject it, and thus my labour will not be lost. I expect this result: so you will not disappoint or annoy me in the least by an adverse decision on the part of the Review" (qtd. in *Letters* 8: 86). Despite Chapman and Eliot's hesitancy about the piece, the article finally appeared in 1854, although Eliot specified that it would be printed in the Independent Section, composed of articles for which the editors refused to accept responsibility (8: 87).[13]

Combe's proposal to systematically categorize individuals to weed out undesirable types was not limited to the prevention of crime. In the fifth chapter of his *Constitution*, Combe recommends that, for the sake of future generations, the constitutionally enfeebled should remain celibate while the healthy should consciously choose partners whose intellectual and moral qualities would ensure the propagation of more perfect offspring (169, 178–79). Such progeny, he explains, will ultimately bring "blessings to the race" by incrementally improving the overall constitution of its character through hereditary transmission (171). Although he admits that selective breeding can improve mental qualities within a race, he cautions that Europeans mating with any other race only dilutes the innate intellectual superiority of the former (171–72). He cites as evidence of this intellectual disparity the "great deficiencies in the moral and intellectual organs" of non-white races, clearly visible on their skulls (166).

Combe's position on "natural" hierarchies was typical of ethnologically minded phrenologists who similarly predicted the possible intellectual achievements of racial groups based on the size and shape of their skulls. Cornelius Donovan, principal of the London Phrenological Institution and the phrenologist from whom Eliot and Bray received a private lesson in 1844, offers an example of this sort of intelligence ranking in his lavishly illustrated *Coombs' New Phrenological Chart* (Figure 5.2).[14]

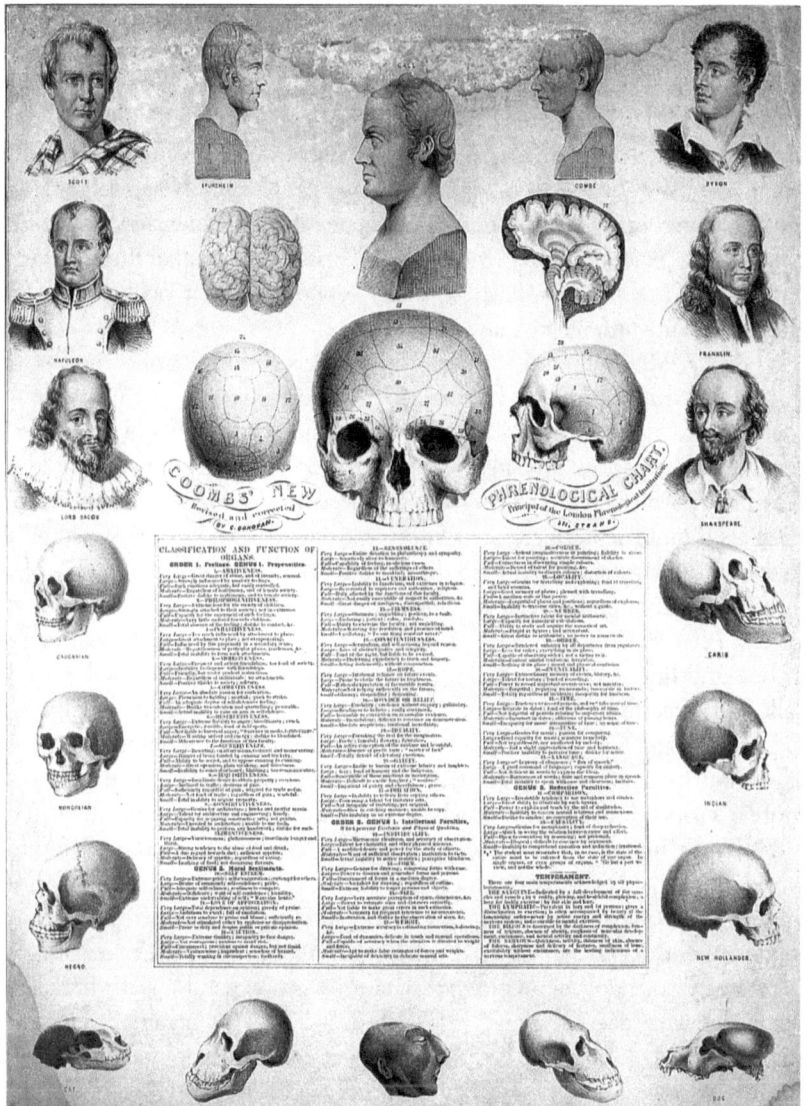

Figure 5.2. Phrenological chart with portraits of historical figures and illustrations of skulls exhibiting racial characteristics. Lithograph by G.E. Madeley, authored by Cornelius Donovan, c. 1850, The Wellcome Collection.

Franz Gall sits in state at center top, flanked by the heads of notable white men: Lord Bacon, Napoleon, Scott, Byron, Benjamin Franklin, and Shakespeare. The representations lose specificity in the lower half of the chart, where skulls representing the various races keep company with the skull of a cat, orangutan, baboon, dog, and (center bottom) the head of an "idiot." Without faces or names, the skulls labeled "New Hollander," "Mongolian," and "Negro" represent types rather than individuals, and the visually evident limits of each race's intellectual capabilities. Elongated, small, or over-large, these skulls supposedly offer ocular proof of the European's innate racial superiority.

In "A Minor Prophet," which opens George Eliot's collection *"The Legend of Jubal" and Other Poems* (1878), Eliot took aim at this phrenological tendency to reify hierarchal distinctions through physical signs of difference, particularly in the case of utopian projections. In it, the reformer Butterworth, armed with the "testimony of his frontal lobe," predicts a biologically perfected future race filled with men possessing "a lobe anterior strong enough / To think away the sand-storms" (33, 73–74). To achieve this perfection, he hopes to eliminate "every faulty human type" (129) in order to create a future in which "All will be harmony of hue and line, / Bodies and minds all perfect, limbs well-turned, / And talk quite free from aught erroneous" (131–33). This scheme, much like Combe's, assumes an unambiguous correspondence between phrenological signs and internal predispositions that then allows Butterworth to promote a vision of the future in which alleged perfection trumps diversity.

Eliot's narrator, however, counters this plan by proposing a future world founded on inclusion rather than exclusion, where richness, through variety, is all. She explains that she "rest[s] in faith / That man's perfection is the crowning flower, / Toward which the urgent sap in life's great tree / Is pressing" so that "the world's great morrows to expand / With broadest petal and with deepest glow" (217–20, 221–22). This organic metaphor stresses a faith in nature's ability to perfect itself, rather than investing in an exclusionary ideology that artificially "constructs / A fairer type" (280–81). To exclude those who seem "hid in harsh forms," is paradoxically to hinder the development of a truly perfect world, enriched by the variety those "types" bring (192). In the poem, Eliot draws a distinction between Butterworth's project of perfecting humankind through exclusionary breeding, and her own, which she believes can be achieved through the more inclusive aims of art, love, and fraternity. Instead of physical assessments, Eliot's future hope rests on affective experiences, including "swellings of the heart and tears" (302), the "labours of the master-artist's hand" (304), "moments of heroic love" (307), and the "conscious triumph of the good within" (309). Ultimately, Eliot reveals Butterworth to be a "minor" prophet because his vision of the future represents only a narrow conception of human perfection, achieved through classification, exclusion, and forced extinction. Unsurprisingly, he projects a world populated with men who resemble himself, each possessing large foreheads. Against this vision of sameness multiplied, Eliot closes the poem with her own prophesy, in which a "Presentiment of better things on earth / Sweeps in with every force that stirs our souls / To admiration, self-renouncing love, / Or thoughts, like light, that bind the world in one . . ." (316–19). These final lines echo the epigraph to "The Lifted Veil," in which the only "light" the poet wishes to receive is that which leads to "energy of human fellowship." Eliot aligns "light," in both cases, with a vision of the future that is inclusive rather than exclusive—rich in variation rather than replete with similitude.

Butterworth serves as a hinge figure in Eliot's evolving argument concerning physiology's relationship to future knowledge. Like Latimer, Butterworth accepts that the body's materiality indexes destiny, and like the visionary character of Mordecai in *Daniel Deronda*, his comprehensive vision of the future addresses the general welfare of humanity. Eliot's ironic use of the word "Prophet" in the poem's title underscores how his teleological mindset mirrors the predictive structure of Judeo-Christian eschatology, but Butterworth presents his redemptive future in distinctly biological terms. Positioning himself as nature's editor, he advocates eliminating "nature's blunders" from future generations (175). In contrast, Eliot argues that there are "lovely minds / Hid in harsh forms—not penetrating them" but "encaged / Like a sweet child within some thick-walled cell" (191–92, 195–96). By predicating his prophesy on a specific conception of phrenologically indexed human perfection, Butterworth encourages an impoverished future that preemptively eliminates latent diversity. By criticizing a tendency to envision the future through biological determinism, Eliot exposes the limitations of epistemologies that, far from encouraging humanity to develop in beneficial ways, might lead to a future that would destroy the variation that makes possible a greater biological, cultural, and moral inheritance.

In *Daniel Deronda*, Eliot's interest in the issues explored in "The Lifted Veil" and "A Minor Prophet" resurfaces as a response to her former arguments about the circumscribing effects of biological determinism. Such a return might be expected since *Deronda* is very much a novel about envisioning the future, with Mordecai assuming the role of the mature visionary in contrast to the despondent Latimer and the satirized Butterworth.[15] Latimer and Mordecai, however, are more particularly connected through the Jewish imagery that enigmatically appears at the midpoint of "The Lifted Veil," just before Latimer confirms his first "wonderfully distinct vision" of the city of Prague from a bridge (9). Before allowing himself to confirm his prevision, Latimer visits a synagogue in the Jewish quarter and recalls, "while our Jewish cicerone reached down the Book of the Law, and read to us in its ancient tongue,—I felt a shuddering impression that this strange building, with its shrunken lights, this surviving withered remnant of medieval Judaism, was of a piece with my vision" (22). After leaving, Latimer rushes to a bridge and finds his mental picture confirmed, "minute in its distinctness down to a patch of rainbow light on the pavement, transmitted through a coloured lamp in the shape of a star" (9). As Brenda McKay has pointed out, the detail references three significant Judaic symbols mentioned in Hebraic literature, all of which are linked to the spiritual gift of prophesy: God's promise to Noah not to flood the world (the rainbow), the promise of a future Messiah (the star), and the Jewish people's future role as a light to the world (the lamp) (78–84). A shared investment in nonlinear temporality links Latimer to Judaism in this scene as he witnesses an ancient faith transported into the present, and likens this with his own experience of seeing something before he has seen it with his actual eyes.

Mordecai's vision is similarly fulfilled on a bridge overlooking a city, where he experiences "with a recovery of impressions that made him quiver as with a presentiment, till at last the nearing figure [of Deronda] lifted up its face towards him—the face of his visions . . ." (422). The confirmation of Latimer and Mordecai's respective visions, however, affects each differently. After trying to avoid a view of the city in order to give himself another day's "suspense," the "only form in which a fearful spirit knows the solace of hope," Latimer rushes to the bridge with a sense of fear and dread (22). Mordecai, however, waits on the bridge with "spiritual eagerness" (422). Whereas for Latimer, the "solace of hope" rests in his vision not being realized, for Mordecai, the recognition of Deronda from the bridge is hope writ large, a signal that he has "'come in time'" to carry on the prophet's work (425). Perhaps the most salient difference between the two rests in the worldview each adopts in response to what their respective visions portend. Unlike Mordecai, Latimer withdraws from the world and isolates himself, passively accepting an inevitable future. This element of the story, in fact, was the moral for George Henry Lewes, who described "The Lifted Veil" as a tale about "the one-sided knowing of things in relation to the self—not whole knowledge" (*Letters* 9.220). In contrast, Mordecai makes clear to Deronda that his visions catalyze connections between himself and others: "'I see, I measure the world as it is, which the vision will create anew. You are not listening to one who raves aloof from the lives of his fellows'" (426). For Latimer, dread, passivity, and solipsism accompany foreknowledge; for Mordecai, eagerness, activity, and communal knowledge attend the experience.

Eliot also makes an important distinction between the formal qualities of the visions. Latimer describes three visions of the future, each conveyed to him through flashes of sight. When the image of Prague breaks upon him, he describes it as a "wonderfully distinct vision—minute in its distinctness down to a patch of rainbow light on the pavement . . ." (9). Similarly clear and specific, his second vision reveals a scene composed of his father, their old neighbor Mrs. Filmore, and Bertha. Despite never having met Bertha, Latimer beholds "a tall, slim, willowy figure, with luxuriant blond hair, arranged in cunning braids" whose "features were sharp, the pale grey eyes at once acute, restless, and sarcastic" (11, 12). In his third vision, Bertha appears to him as his future wife, entering his father's library with a candle and standing before the hearth "with cruel eyes, with green jewels and green leaves on her white ball-dress" and a "great emerald brooch on her bosom, a studded serpent with diamond eyes" (19–20). The visual elements of this final image are so dramatically rendered that Latimer "wakes" from the experience with his "eyelids quivering" and "the green serpent with the diamond eyes remaining a dark image on [his] retina" (20). Taken together, these visions are so vivid and clear that Latimer can particularize everything from a twist of hair to a patch of light on the ground. He describes the scenes as if he were studying a sharply rendered photograph, forever present in his mind's eye. His experience of second sight is literal: he sees exactly what he actually will see in the future with his physical eyes.

This experience of the future, which Latimer refers to as "prevision" (14, 18) contrasts with Mordecai's, which is more akin to the romantic visionary experience. As Jay Clayton has pointed out, a "loss of sight" is a key feature of romantic transcendence (*Romantic Vision* 8). For Wordsworth, the visionary poet George Eliot most frequently quoted, "greatness" abided in those moments "when the light of sense / Goes out, but with a flash that has revealed / The invisible world" (*The Prelude* 6.600–602). The mark of an authentic vision is "that bodily eyes / Were utterly forgotten, and what I saw / Appeared like something in myself, a dream, / A prospect in the mind" (2.349–52). While the poet "sees" something, it is numinous, inward, and an extension of the self, rather than something exteriorized and distinctly material (that seen with "bodily eyes"). This experience resembles Mordecai's vision of the one he has long awaited:

> His [Mordecai's] inward need for the conception of this expanded, prolonged self was reflected as an outward necessity . . . his coherent trains of thought often resembled . . . genuine dreams in their way of breaking off the passage from the known to the unknown. Thus, for a long while, he habitually thought of the Being answering to his need as one distantly approaching or turning his back towards him, darkly painted against a golden sky. [. . .] Thus it happened that the figure representative of Mordecai's longing was mentally seen darkened by the excess of light in the aerial background. (406)

Mordecai never distinctly apprehends Deronda in his vision, except as a darkened, indistinct shape in a beautiful setting. Further, Mordecai can only conjure the image as a dreamlike extension of himself, not unlike Wordsworth's visionary experience at the Simplon Pass, which "Appeared like something in myself, a dream." As Mary Wilson Carpenter observes, "the vision has arisen because of Mordecai's constitution and the sphere of ideas in which he moves, and none of the elements in it are new. They are in fact a clearly recognizable example of poetic or artistic vision . . ." (143). Lacking specificity, Mordecai's vision emerges as the indistinct image of a concept that has yet to manifest itself. Like poetic inspiration, he views Deronda as the fulfillment of an idea that will connect him with a larger view; he is the "flash that has revealed / The invisible world" through his very visibility (Wordsworth 6: 601–2).

That said, Mordecai's recognition of Deronda is importantly tied to the latter's visual distinctiveness. Long before meeting Deronda, Mordecai, being "Sensitive to physical characteristics" had, in England and abroad, "looked at pictures as well as men" and had "sometimes lingered in the National Gallery in search of paintings which might feed his hopefulness with grave and noble types of the human form, such as might well belong to men of his own race" (405). In his vision, these many faces blend, and "the words youth, beauty, refinement, Jewish birth, noble gravity, turned into hardly individual but typical form and colour: gathered from his memory of faces seen among the Jews of Holland and Bohemia, and from the paintings which revived that memory" (407). As Daniel Novak has argued, Mordecai's method of indistinct yet suggestive identification anticipates the process Francis Galton employed to produce composite photographs of racial, criminal, and familial "types" just two years after the publication of *Daniel Deronda*.[16] In the same way that a composite photograph can singularly represent the characteristics of a class, Mordecai views Deronda's face as a particular instantiation of a generalized image of Jewish perfection produced through the superimposition of noble Jewish faces in his mind.

As Eliot demonstrates in her criticism of other "visionaries," however, imagining the future through physical typing is risky. Composite photographs essentially operate as compressed images of the physiological "types" that phrenologists like Combe sought to identify through compulsory examinations and skull classification. Unsurprisingly, just as Combe recommended the identification of potential criminals through their small anterior lobes, Galton suggested that composite images of the "criminal type" could be used to identify "the man who is liable to fall into crime" (*Inquiries* 363). Although Galton's composite photography postdates *Deronda*, the habits of mind that informed his agenda were precisely those that shaped the phrenological aims Eliot criticized nearly two decades before Galton would coin the term "eugenics."[17] By allowing an abstract conception of "the typical" to take precedence over the particular, Galton, Combe, and Deville—much like the fictional phrenologists Letherall and Butterworth—all adopt a similar way of seeing into the future.

Although Mordecai appears to engage in a similar process of visual abstraction when identifying Deronda, Eliot draws a crucial distinction between Mordecai and her earlier visionaries in the opening paragraph of the chapter containing Mordecai's recognition of Deronda from Blackfriar's bridge. She writes:

> "Second-sight" is a flag over disputed ground. But it is matter of knowledge that there are persons whose yearnings [and] conceptions . . . take the form of images which have a fore-

> shadowing power: the deed they would do starts up before them in complete shape, making a coercive type; the event they hunger for or dread rises into vision with a seed-like growth, feeding itself fast on unnumbered impressions. . . . No doubt there are abject specimens of the visionary . . . but what great mental or social type is free from specimens whose insignificance is both ugly and noxious? (404)

Eliot's reference to those who see "the deed they would do" prefigures Gwendolen's early vision of the drowning man, later Grandcourt, who she refuses to save—but the passage also recalls Latimer, who, much more than Gwendolen, belongs to the "mental or social type" that experiences second sight routinely. Latimer also resembles one of the "abject specimens" of the visionary, whose perceived "insignificance" is the theme of his own solipsistic story, and whose visions lead him to become more reclusive and unconnected to the outside world. For Mordecai, however, the experience of the vision never divorces itself from a hermeneutic endeavor. While his vision of Deronda is literal (he sees a man on a boat from a bridge, illuminated from behind by a golden sky), a cognitive process intervenes in his experience of the vision. It becomes the outgrowth of a "mature spiritual need" for an "ideal life straining to embody itself," and as the literal vision advances in his mind, he supplies the concepts that will define what he inwardly yearns for in "a companion and auditor" (407). The image of Deronda bears the stamp of Mordecai's future hope—the role of visionary, for him, is active, and requires a reworking of the physical text preternaturally provided.

Although likened to a "type," Deronda's resemblance to idealized images opens rather than forecloses the possibilities of Jewish achievement. This distinction is akin to the difference Carlo Ginzburg has pointed out between Galton and Wittgenstein's interpretations of the former's composite photography. As aforementioned, Galton believed the purpose of these generalized portraits lay in identifying aberrant individuals through their visual likeness to the generic image. Like those individuals with small anterior lobes whom Combe wished to have incarcerated before committing crimes, Galton's image of the potential criminal serves as a visual pattern for identifying undesirables through their physical appearance. Disregarding Galton's intention, Wittgenstein focused on how these superimposed images represented a new way of thinking about the individual. According to Ginzburg, Wittgenstein used the composite portraits in a way Galton never intended—to help him "articulate a new notion of the individual: flexible, blurred, open-ended, like the elusive characters we come across in Proust's novel" (549). This use of the type to "articulate a new notion of the individual" is akin to what Eliot attempts through Mordecai's visions. The person who will fulfill the vision must be "flexible, blurred, open-ended"—in Deronda's case, this means a willingness to accept the cultural inheritance of his people transmitted through Mordecai. Mordecai desires "some young ear into which he could pour his mind as a testament, some soul kindred enough to accept the spiritual product of his own brief, painful life, as a mission to be executed" (404). Deronda must be both receptive and sympathetic, qualities also useful for a future in which his own desires will be inextricably linked with the welfare of his people. In a sense, Deronda must be open at both ends: open to receive Mordecai's teachings in the present, and open to enlarging the "possibilities of the Jew" in the future.

While the process of true visionary identification requires an abstract mental attitude, Eliot suggests that what the visionary apprehends must also have a prepossessing power that is visible to all. A considerable amount of scholarship has focused on the vagueness of the narrator's descriptions of Deronda, which render him strangely inaccessible: George Levine claims he is "alien and hence potentially beyond the limits of common desires" (20), Bryan Cheyette remarks that the "'invisibility' of Deronda's vocation renders his character passive and featureless" (45), and U.C. Knoepflmacher observes that "his characterization asks . . . for a receptiveness which can be accorded neither by the religious devotee nor the skeptical critic of art" (147). Extending the lack of definition to include Mirah and Mordecai as well, Henry James writes that they "produce no illusion. They are described and analyzed to death . . . but that doesn't make real figures of them. They have no existence outside of the author's study" (686). Citing these same sources, Peter Capuano remarks that "Critics of virtually every stripe since James have used Deronda's physical abstraction as the cornerstone for a myriad of disparate interpretations" (155), a comment that echoes Daniel Novak's observation that "since the novel's publication, critics have complained or conceded that Deronda is merely a disembodied type" (109). The critics to which

Novak and Capuano refer, however, are addressing Deronda's characterization rather than how Eliot represents his physical presence and visual impressiveness. In fact, Knoepflmacher draws attention to Eliot's tendency to compare Deronda to the subjects in Italian portraiture, and Levine explicitly states that "George Eliot carefully makes him striking in appearance . . ." (20). To extend Eliot's ambiguous delineation of Deronda's personality to his physical presence effaces an important aspect of his political and social usefulness, which Mordecai knows will require both visual distinctiveness and an indistinct (and thus impressible) character.

In fact, Eliot repeatedly and emphatically asserts that much of the interest Deronda inspires in others comes from his appearance. Compared to the portraits of Sir Hugo's ancestors that line the gallery walls, Deronda is "handsomer than any of them, and when he was thirteen might have served as model for any painter who wanted to image the most memorable of boys: you could hardly have seen his face thoroughly meeting yours without believing that human creatures had done nobly in times past, and might do more nobly in time to come" (141). Notably it is Deronda's face, in his bar mitzvah year, that can make people believe in the nobility of past and future generations. His affecting visual presence acts as a promissory note of future achievement guaranteed by the biological mechanism of inheritance. As an adult, this power only intensifies and manifests itself in every part of his body and mien:

> Rowing in his dark-blue shirt and skull-cap, his curls closely clipped, his mouth beset with abundant soft waves of beard, he bore only disguised traces of the seraphic boy . . . only to look at his lithe powerful frame and the firm gravity of his face would have been enough for an experienced guess that he had no rare and ravishing tenor such as nature reluctantly makes at some sacrifice. Look at his hands . . . they are long, flexible, firmly-grasping hands, such as Titian has painted in a picture where he wanted to show the combination of refinement with force. And there is something of a likeness, too, between the faces belonging to the hands—in both the uniform pale-brown skin, the perpendicular brow, the calmly penetrating eyes. Not seraphic any longer: thoroughly terrestrial and manly; but still of a kind to raise belief in a human dignity . . . often the grand meaning of faces as well as of written words may lie chiefly in the impressions of those who

look on them. But it is precisely such impressions that happen just now to be of importance in relation to Deronda . . . passing under Kew Bridge with no thought of an adventure in which his appearance was likely to play any part. In fact, he objected very strongly to the notion, which others had not allowed him to escape, that his appearance was of a kind to draw attention. (157–58)

The narrator's assertion that the significance of a face is proportional to the degree that it makes an impression distinguishes Eliot's strategic use of the body's materiality from the biological determinism of her earlier phrenological visionaries. A face does not have a "grand meaning" in and of itself but in how it affects others, suggesting that it potentially has a rhetorical force similar to writing, but that also depends on the degree to which others imbue it with significance. The power of Deronda's physical presence is as much a product of cultural transmission as biological inheritance because it stems from his resemblance to the subjects in great works of art that have already associated such physical traits with the qualities of "refinement," "force," and "dignity." In a sense, Deronda is a work of art in flesh and blood, a fact Mordecai can later recognize in part because he has long sought out paintings of "young, grand, and beautiful" men in the National Gallery (405). In this key scene, Eliot explicitly states that Deronda's "appearance" plays a significant role in his hero's journey. Occurring right before his first encounter and rescue of Mirah, "appearance" may carry with it the secondary meaning of a fortuitous arrival at a crucial moment. In this spatiotemporal sense, so critical for a plot that turns on coincidences that might be read as destiny, the scene also foreshadows Mordecai's recognition of Deronda from Blackfriar's bridge. Syntactically, however, "appearance" directly refers to his visible presence, the beauty and impressiveness of which the narrator has spent a lengthy paragraph detailing.

Deronda's conspicuous physicality allows Mordecai to advance a future plan that eschews two dangers inherent in visualizing race, the first being a dominant culture's tendency to render a disempowered group invisible through assimilation. In the conversation among the Jewish philosophers at the Hand and Banner, Gideon advocates just this possibility as a form of racial progress: "'There's no reason now why we shouldn't melt gradually into the populations we live among. That's the order of the day in point of progress. I would as soon my children married

Christians as Jews'" (449–50). This sentiment echoes Mrs. Meyrick's earlier observation that "'if Jews and Jewesses went on changing their religion, and making no difference between themselves and Christians, there would come a time when there would be no Jews *to be seen*'" (317, my emphasis). While this rhetoric of assimilation seems to advocate acceptance, the ultimate aim is to render Jews invisible.[18] At the other end of this spectrum rests the danger of becoming conspicuous through negative stereotypes. Sander Gilman explains this cultural process in his description of the icon, noting that "Specific individual realities" are often "given mythic extension through association with the qualities of a class. These realities manifest as icons representing perceived attributes of the class into which the individual has been placed. The myths associated with the class, the myth of difference from the rest of humanity, is thus, to an extent, composed of fragments of the real world, perceived through the ideological bias of the observer" ("Black Bodies, White Bodies" 204). Representational icons created by the dominant class to describe another class, as Gilman explains, are never objective but rather are always linked to an "ideological bias" that affects how the observed class is perceived. The resulting stereotypes, as Gilman further demonstrates in *The Jew's Body*, can attach exaggerated, negative, and/or pathologized attributes to the observed class through visual images. The "myth of difference" informs Deronda's first imagining of Mirah's mother and brother: "Deronda's thinking went on in rapid images of what might be: he saw himself guided by some official scout into a dingy street . . . and saw a hawk-eyed woman, rough-headed, and unwashed, cheapening a hungry girl's last bit of finery; or . . . under the breath of a young Jew talkative and familiar, willing to show his acquaintance with gentlemen's taste" (176–77). Eliot explains that these images are the product of Deronda's lack of experience with "actual Jews," which leaves him to piece together his impressions of the race through "ugly stories of Jewish characteristics and occupations" which he converts into a mental montage of ugly images (176). At this point in the novel, Mirah seems an anomaly, for "the facts he knew about [Jews] were chiefly of the sort most repugnant to him," and so he wishes to save her from "an association with what was hateful or contaminating" (176). Early on, Deronda can only envision Jewishness through the lens of racial stereotypes: to be Jewish is to live on a "dingy street," to be "unwashed," and unctuously verbose—to be "contaminating" and hated. Hans also advances a stereotypical image in an offhand joke about Jewish physicality, observing, "'Every male of that race is insupportable,—'insupportably advancing'—his nose' " (396). Ultimately, these iconographic images of Jewishness take on a reality of their own, with powerfully detrimental effects.

In fact, one of the most formative moments of Mirah's life was having overheard a comment not unlike Hans's:

> One day, when I was looking at the sea and nobody took notice of me, I overheard a gentleman say, "Oh, he is one of those clever Jews—a rascal, I shouldn't wonder. There's no race like them for cunning in the men and beauty in the women. I wonder what market he means that daughter for." When I heard this, it darted into my mind that the unhappiness in my life came from my being a Jewess, and that always, to the end the world would think slightly of me and that I must bear it, for I should be judged by that name. (183)

In the encounter, Mirah and her father are merely seen but are nevertheless quickly categorized and stigmatized through generalized assumptions about their race. Mirah realizes that she will always first "be judged" not as an individual but as a Jewess, with all of the negative connotations the Western world has heaped upon "that name." After learning that the race to which she belongs will always take precedence over her individual attributes, Mirah reconciles herself to finding comfort in the belief that her "suffering was part of the affliction of [her] people, [her] part in the long song of mourning that has been going on through ages and ages" (183). This is one reaction to social typing—finding solidarity through suffering—that does not allow the bias of the dominant group to persuade the minority into assimilation. Making the suffering and stigmatization of her people her own, Mirah insists, " 'I will always be a Jewess. I will love Christians when they are good. . . . But I will always cling to my people' " (317). While open to associating with others, Mirah, much like Mordecai, wishes to remain anchored to that which makes her different from Christians rather than attempting to blend into the population in the way Gideon suggests.

The novel's treatment of the dominant culture's power to bring about the self-imposed erasure of Jewish identity culminates in the Princess's revelation that she decided to give her son the "advantage" of being reared and recognized as a Gentile. Like Mirah, she completely accepts that Jewishness will always be the trait by which she will be publicly recognized, but unlike her, she does not find solace in the solidarity of shared oppression and so instead turns to a strategy of passing. For the Princess, to acknowledge cultural distinction is to assume a curse, explaining, "'I rid myself of the Jewish tatters and gibberish that make people nudge each other at sight of us, as if we were tattooed under our clothes, though our faces are as whole as theirs. I delivered you from the pelting contempt that pursues Jewish separateness'" (544). Tellingly, she likens being Jewish in European society to having a body believed to be marked with indelible ink, a distinction mentally written upon the Jew's body when seen, even though it is purely fictive. For her, this misguided perception might as well be true, because to be recognizably Jewish in such a society is to assume the mantle of oppression. Like most of the "philosophers" at the Hand and Banner, she believes that the Jewish population has become too visible as a stigmatized class and accepts that the rational solution is to make as little distinction as possible between Jewish and Christian identity in order to avoid oppression. In this way, racial visibility renders a group ideologically invisible, resulting in the inevitable extinction of a cultural inheritance that society stigmatizes as a Jewish tendency toward "superstitions and exclusiveness" (449).[19] Throughout the novel, the representations of the race from outside the Jewish community—Deronda's early association of Jewishness with dirtiness, the man who claims the race is known for "beauty in women," Hans's racist joke about noses—take the form of stigmatized *visual* images. The ultimate result of this embedded cultural perception by the dominant group is a concession by the oppressed class that while the iconic representation is misrepresentative, it has a collective, coercive power.

In response to this problem, Mordecai assumes the position of an author who determines the body's material destiny by imbuing it with meaning. As Rosemarie Bodenheimer has pointed out, the very idea of authorship was a vexed one for Eliot because an author occupies a position of authority that necessarily conflicted with her desire to suppress ambition and egoism. To balance these opposed states, in her letters she casts herself "as a sufferer rather than as one of the ambitious, authoritative narrators of her books" (164). This perhaps explains why Eliot counterbalances the ability to both foresee and determine Deronda's destiny with Mordecai's extreme physical suffering. Tellingly, at the very moment Deronda remembers Mordecai's insistent command, "'You must hope my hopes—see the vision I point to—behold a glory where I behold it,'" he sees the visionary as a "suffering reality . . . a man steeped in poverty and obscurity, weakened by disease, consciously within the shadow of advancing death" (455). Paradoxically, the price Mordecai must pay for his narrative authority renders him unable to embody his own vision. When he describes to Deronda his past attempts to revive a sense of connectivity among his race, he reveals, "'I found none to listen. . . . No wonder. I looked poor . . .'" (427). He explains to Deronda that even writing could not communicate his message, likening his would-be texts to his own frail body: "New writing of mine would be like this body . . . within it there might be the Ruach-ha-kodesh—the breath of divine thought—but men would smile at it and say, 'A poor Jew!'—and the chief smilers would be of my own people" (428). Because his physical body and visible presence negates the power of his message before he has even attempted to deliver it, Mordecai searches for a new body that will outwardly serve as "the more beautiful, the stronger, the more-executive self"—one who shares his beliefs, but also who, through his physical presence, announces the ability to manifest these ideas. Deronda's body provides Mordecai with the "terrestrial" material that can give physical shape to his own abstract ideas—someone capable of action but also one whose corporeal presence resists the public's tendency to classify and dismiss individual representatives of a race based on immediate visual impressions (406).

As a rhetorical strategy, Mordecai's use of Deronda's prepossessing corporality might best be understood as creating an emblem—a visual instantiation of his "firmest theoretic convictions" in a "specific self-asserting form" (411). Like the other visually typifying forms of Jewish identity addressed in this chapter, the emblem locates the visual value of an image in its ability to stand for more abstract or general concepts. However, whereas collective ideological bias produces representative racial iconography, an agent or author intentionally creates an emblem for a specific purpose.[20] In other words, whereas the icon results from the repeated act of being claimed by the dominant culture, an author actively creates an emblem by claiming a powerful image and endowing it with ideological significance. Full accepting as inextricable the link between the envisioned future and the embodied subject, Mordecai finally uses his "[s]ensitiv[ity] to physical characteristics" to identify "an embodiment unlike his own," which, after being invested with his beliefs, can outwardly "glorify the possibilities of the Jew" (405).

In many ways, Mordecai's plan resembles the interventionist tactic Gayatri Spivak has identified as strategic essentialism, or the "*strategic* use of positivist essentialism in a scrupulously visible political interest" in order to escape the historical invisibility that results from discriminatory practices (205). While the effort to recover the peasant or "subaltern consciousness" is vulnerable to postmodern critique as an illusory search for a centering presence, it still has value as a political act, representing "the always asymmetrical relationship between the interpretation and transformation of the world . . ." (*In Other Worlds* 208). In her reading of Spivak, Diana Fuss argues that essentialism, in this instance, emerges as viable and worthwhile because it is deployed in the service of a disempowered group. Thus, the "permissibility" of essentialism is ultimately "framed and determined by the subject-position from which one speaks" (32). This ethical distinction between permissible and impermissible uses of essentialism also clarifies the difference between a discriminatory essentialism that produces Jewish iconography and Mordecai's clearly strategic attempt to "[r]evive the organic centre" of his people through Deronda's visual distinction (454).

Deronda's characteristic receptivity, or "keenly perceptive sympathetic emotiveness," renders him particularly well suited for the role of living emblem (425).[21] This quality allows Deronda to not only "receive from Mordecai's mind the complete ideal shape" of "personal duty and citizenship" (437) but also to receive the sympathetic investment of others. In her study of George Eliot's use of nineteenth-century organic theory, Sally Shuttleworth identifies Deronda's "sense of obligation with directed energy" as a representation of the "organic ideal," enabling him to integrate his individual identity within a larger social function (*George Eliot* 191). For Shuttleworth, this power is physiological insofar as it is psychological, but Eliot extends this integrative force from the interior to the exterior—from the physiological psychology of the mind to visual cues that invite interpersonal connection. As Eliot observes, "Receptiveness is a rare and massive power," and it is this quality that "gave Deronda's face its utmost expression of calm benignant force" (425). Significantly, Deronda's *face* wields this power because his striking features lead others to automatically attribute to him an investment in their own welfare. Mordecai, for instance, finds himself able to tell Deronda of his future plan because the "receptiveness" of his expression "nourished Mordecai's confidence and made an open way before him" (425). Similarly, Gwendolen confides in Deronda because his "eyes had a peculiarity . . . which seemed to express a special interest in every one on whom he fixed them" (280). Deronda, then, visibly represents one of Eliot's most cherished characteristics—sympathy—but in an executive form that dislodges the quality from its passive connotations by externalizing its power.

As a receptive corporeal vessel, Deronda's physical presence allows Mordecai to reclaim signifying power through the Jewish body. Even before meeting Deronda, Mordecai was attempting the process of ideological authorship through his Hebrew lessons with Jacob, likening the act to inscribing the body's interior, as he thinks, "'The boy will get them engraved within him . . . it is a way of printing'" (408). Fittingly, at the very moment Mordecai decides that his efforts with Jacob will prove fruitless, he is "struck by the appearance of Deronda" in the bookstore (410). Faced with the pressing concern of how to transform ideas into a political reality, Mordecai begins printing on the more receptive tablet of Deronda, into whom he can "pour his mind as a testament," imprinting "the spiritual product of his own brief, painful life, as a mission to be executed" (404). Seemingly by the sheer force of Mordecai's convictions Deronda accepts this role, revealing that Mordecai's visionary power lies less in an otherworldly ability to predict the future and more in actively creating that future by casting the role of emblematic leader and providing Deronda with a script. Resigned to the fact that bodies *will* signify in limiting and troubling ways, Mordecai pragmatically considers his audience and finds the body best equipped to fulfill his vision.

Importantly, Deronda's physical body is always connected, in Mordecai's mind, with the future of the body politic of Israel. At the Hand and Banner, he explains that the only hope for the future of Judaism is to "Revive the organic centre: let the unity of Israel which has made the growth and form of its religion be an outward reality" (454). Abstract ideas and the spirit of Judaism alone, he posits, cannot thrive divorced from something that can be located and seen. Diffusion and disconnection from an "outward reality" is what threatens to extinguish the immaterial, spiritual, and intellectual nature of Judaism by uprooting it from a material presence in the world. Mordecai likens this state to a loss of sight, explaining that his people "'have no vision; in their darkness they are unable to divine.'" Mordecai's fervent speech about the necessity of an outward, visible center for the Jews is fittingly inspired by, and directed to, Deronda: "His extraordinary excitement was certainly due to Deronda's presence: it was to Deronda that he was speaking . . ." (454). The relationship between Mordecai and Deronda enacts on a small scale the visionary's larger hope: just as Deronda embodies Mordecai's inward vision, Israel will become the national body of Jewish consciousness, with "a heart and brain to watch and guide and execute" (456). Without the literal embodiment of a revived feeling of nationality among the Jews (represented through Deronda's fulfillment of his prophesy), Mordecai's future hope would merely be an idea without any transformative power. Deronda's body, and the "organic centre" of nationality it stands for and will manifest is the material Mordecai invests with a new significance. Accepting that the perception of racial identity has real effects, he rejects following Pash, Gideon, and Leonora to their "rational" conclusion that iconic representations of Jewishness necessitates cultural assimilation. Rather than attempting to dismantle the racial iconography that distorts social perception, Mordecai selects a paragon upon which to fasten his vision. Deronda's conspicuous presence emblematizes the separation and distinction that will paradoxically increase the possibilities for greater sympathy among nations.

In this way, Mordecai offers an alternative to the problems that attend a consideration of future knowledge and the body's legibility. Similar to Latimer's passivity in the wake of a foreordained destiny, an attitude of "rational" resignation accompanies the acceptance of cultural assimilation in *Daniel Deronda*. Similarly, the pervasive ideological bias that projects and encourages "'a time when there would be no Jews to be seen'" (317) parallels Butterworth's eugenic project to breed out undesirable types. In stark contrast to these perspectives, Mordecai presents Deronda with another vision of the future, explaining, "'In the doctrine of the Cabbala, souls are born again and again in new bodies till they are perfected and purified, and a soul liberated from a worn-out body may join the fellow-soul that needs it, that they may be perfected together, and their earthly work accomplished'" (461). This vision promotes a sort of spiritual eugenics, where "purity" refers to metaphysical, rather than biological, perfection. Rather than winnowing down the range of human possibilities through a future design based on assimilation or artificial selection, this Cabbalistic model favors the concentrated accretion of spiritual inheritance. This plan, however, does not divorce the spiritual from the material in its practicality—indeed, the souls of the "worn-out body" require "new bodies" to accomplish "earthly work." Similarly, Mordecai's mind contains the cultural inheritance that Deronda's apt body receives, and their lives will then be justified through the collective perfection that attends the earthly work of nation-building.

In this way, Eliot's early interrogation of the ideological assumptions informing phrenological logic resurfaces in her argument about the body and future knowledge in *Deronda*. Scientific visionaries such as Combe and Galton relished the possibility of indexing character through material signs as the great hope for the future of humanity because it simplified distinctions and made possible a new organizing structure for society. For both, it was a short leap from recognizing difference to imposing an evaluative hierarchy onto the types they had "objectively" identified through biological signs, which in turn justified projects to minimize difference by purging undesirables from the population. In her critique of biological determinism, Eliot imagines Latimer and Butterworth at opposite ends of a power spectrum underlying the social compulsion to type, classify, and control. Butterworth, as active agent, attempts to justify and reinforce his dominance, whereas Latimer exists as the passive object and victim of such a paradigm. Yet, while offering a critique of the ideological biases that structured this way of thinking through cautionary examples, neither work offers a solution to the problem of envisioning the future through the body. Through Mordecai, Eliot models a form of strategic essentialism that both recognizes the power of presence and yet avoids the delimiting effects of social typing. Mordecai's interpretation and deployment of physicality does not participate in reifying racial iconography but, rather, recasts future possibilities by positing Deronda as one able to change culture through an emblematic presence that is of a piece with an emerging ideal. Thus, the body comes to matter in a more subversive and overtly political way because the future vision that motivates interpretation privileges difference and variation in the development of a more inclusive worldview. This reflects Eliot's own belief that "the human race has not been educated on a plan of uniformity, and it is precisely that partition of mankind into races and nations" that has "been the means of enriching and rendering more and more complete man's knowledge of the inner and outer world" ("A Word for the Germans" 388). For Mordecai, the vision that motivates the interpretation of the body embraces distinction in order to encourage the proliferation of variation. Set apart by Mordecai's inclusive ideology rather than an exclusionary agenda, Deronda serves as a model for the practical achievement of a more comprehensive cultural inheritance.

Much like the other women authors discussed in previous chapters, Eliot ultimately makes use of essentialism for progressive ends, albeit in a way that locates nature's force on the surface of the body rather than in the brain. The value of this strategy, however, loses its ethical dimension when transferred to a context that takes account of gender in addition to race. In defending her decision to abandon Deronda in pursuit of a professional career, the Princess hands him a miniature of herself and asks,

> "Had I not a rightful claim to be something more than a mere daughter and mother?" The voice and the genius matched the face. "Whatever else was wrong, acknowledge that I had a right to be an artist, though my father's will was against it. My nature gave me a charter."
> "I do acknowledge that," said Deronda, looking from the miniature to her face, which even in its warm pallor had an expression of living force beyond anything that the pencil could show." (570)

Much like the feminist phrenologists discussed in the first chapter of this book, Alcharisi uses an appeal to innate capabilities that are physically evident to justify her break from social norms and patriarchal expectations. Her inherent nature provides proof of uncommon aptitudes, which even her son, who has so keenly felt her abandonment, cannot deny. The acknowledgment of her "living force" through a visual appraisal of her face in comparison with an artistic representation recalls the narrator's earlier recognition of Deronda's visual "combination of refinement with force" that becomes articulable through an association with the Titian painting. Eliot makes clear that this innate (and apparently, heritable) greatness has a palpable and irrefutable power, which raises the issue of why this gambit for authority seems permissible in Deronda's case but morally dubious in the case of the Princess. The crux of the distinction seems most obviously tied to their motivations, with Alcharisi seeking personal fulfillment and fame whereas Deronda selflessly dedicates his life to the leadership and service of his people. And yet, despite this apparent similarity, their positions are hardly equivalent. As Alcharisi points out to him, "You are not a woman . . . you can never imagine what it is to have a man's force of genius in you, and yet to suffer the slavery of being a girl" (541). The "force" that Deronda has the agency to exercise in any way he chooses was for his mother a curse insofar as her natural ability foregrounded how her environment was completely incommensurate with her innate abilities, interests, and talents. In a sense, she is doubly imprisoned: first by the expectations of a controlling patriarch, and again by the knowledge of her repressed genius.

Alcharisi's argument has points of similarity with Spivak's skepticism about the ultimate utility of strategic essentialism when sex is taken into account. Spivak explains that because women have not yet been enfolded into the essentialism of kinship in any real way, their difference has typically been used as a sign to consolidate bonds of an essential subaltern identity, and then effaced so a male subject position stands as symbolically coherent: "[T]he figure of the woman, moving from clan to clan, and family to family as daughter/sister and wife/mother, syntaxes patriarchal continuity even as she is herself drained of proper identity. In this particular area, the continuity of community or history, for subaltern and historian alike, is produced on . . . the dissimulation of her discontinuity, on the repeated emptying of her meaning as instrument" (*In Other Worlds* 220). And indeed, Deronda benefits from all the consolidating effects of "patriarchal continuity" by reifying his Jewish identity through his marriage to Mirah. She is the flesh-and-blood counterpart to the spiritual transference bequeathed to him by Mordecai on the latter's deathbed. Conveniently for Deronda, Mirah's "nature" happens to match her role as the supportive Jewish sister/wife/mother. Alcharisi makes sure of this in her interrogation of Deronda regarding his future bride, fittingly communicated through a series of negations:

> "[She is] [n]ot ambitious?"
> "No, I think not."
> "Not one who must have a path of her own?"
> "I think her nature is not given to make great claims."
> "She is not like that?" (569)

It is at this point that she produces the portrait of herself that so indisputably establishes her essentialist "right" to abandon the role of wife and mother to pursue a profession. Mirah emerges as the absence of the identity Alcharisi asserted for herself. As the not-Alcharisi, Mirah balances out the chiasmatic symmetry of the Princess's past and Mirah's future: just as Alcharisi willfully abandoned Deronda to pursue an independent life, Mirah willingly rejects the life of an artist to accept the role of dependent wife. Beyond the mirrored reversal of their respective positions, this inversion also has implications for gendered subjectivity. In a statement that reverses

Simone de Beauvoir's theory of male/female relational dynamics, Alcharisi posits, "I was never willingly subject to any man. Men have been subject to me" (571). In stark contrast, Deronda's acceptance of his biological destiny is finally shored up in relation to Mirah. When he declares, "I consider it my duty . . . to identify myself, as far as possible, with my hereditary people, and if I can see any work to be done for them that I can give my soul and hand to, I shall choose to do it," his mother penetratingly infers, "You are in love with a Jewess" (566, 567). Deronda rejects this explanation, and of course the Princess intends to suggest a more mundane rationale for her son's fervent desire to adhere to his Jewishness. In terms of structural arrangements, however, Deronda reifies the spiritual identity Mordecai outlines for him in an outward, physical, and (assuming progeny) hereditary sense through his union with Mirah. In the final chapter, Mirah fulfills the duty of the good Jewish woman that the Princess shirked: to perform as "an instrument" of the patriarch's designs (567).

This seems to confirm Spivak's suspicion that women will likely remain subjugated under essentialist appeals precisely because of the ubiquity of constitutive patriarchal oppression. This caveat to essentialism's potential subversive utility, however, is only valid if the essentialism to which one appeals is grounded in sexual difference. Alcharisi's essentialism, however, is not cultural or spiritual—it is biological—and she premises it on the natural equality of mind despite one's gender. Her essentialism is not "strategic" but is nevertheless radical and subversive—perhaps too radical for Eliot's pragmatic incrementalism regarding the future progress of sexual equality. Alcharisi's narrative suggests that the force of nature and opportunity can potentially liberate one woman, but not many, because the tactic finds purchase only through exceptionalism. As she frankly admits, the belief in Jewish women's skill in singing and acting fortuitously increases their professional acceptability, making them slightly more likely to achieve independence: "[My father] hated that Jewish women should be thought of by the Christian world as a sort of ware. . . . As if we were not the more enviable for that! That is a chance of escaping from bondage" (541). The distinctiveness that sets both mother and son apart in the same historical moment operates differently in their separate gendered contexts. The remarkable visual presence that ties Deronda to a shared heritage and larger community simply marks his mother as an exotic spectacle. Paradoxically, the only way Alcharisi could release herself from the bondage of her sex is by taking advantage of racial stereotypes. While Deronda views the choices she has made as a betrayal of her race, the inverse is also true: to honor her hereditary "duty" as a Jewish woman is to betray her natural right to freedom from the domestic sphere.

Ultimately, the novel appears to heuristically endorse an appeal to an essential Jewish identity as a way to galvanize an oppressed group, while also acknowledging that women are circumscribed even within movements that advocate for the recognition of human equality. The Princess's arguments rest uneasily alongside the inclusive optimism to which Deronda clings. As she explains to her son, to accept such an inheritance for a woman is "[t]o have a pattern cut out [and be told] 'this is the Jewish woman; this is what you must be; this is what you are wanted for; a woman's heart must be of such a size and no larger, else it must be pressed small . . ." (541). This critique suggests that Mordecai's vision of a unified, inclusive future state that promises to enfold the many still has room for only one manner of woman. The irreconcilability of Alcharisi's position with Mordecai's hopeful expectations exposes a problem the novel represents but does not solve—that if a woman's innate capacities are at odds with her social role, then the expectations that bind her are unjust. After all, it is Alcharisi's acute awareness of her natural abilities that justifies her rebellion as ethical: "I wanted to live out the life that was in me. . . . I had a right to be free. I had a right to seek my freedom from a bondage that I hated" (536–37). Her argument that innate dispositions justify her share in a larger life is the same as that invoked by female phrenologists who advocated for women's rights on the scientific basis of intellectual equality. While her motives are self-serving and antithetical to Eliot's cherished values of sympathy, sacrifice, and compassion, the narrative nevertheless endorses the validity of her justification: "Whatever else was wrong . . . I had a right to be an artist, though my father's will was against it. My nature gave me a charter" (570). In Eliot's fiction, this is one of the most radical statements relating to gender politics. While the novel does not further pursue the idea, it is nevertheless a powerful articulation of women's circumscribed lives in light of their mental capabilities, written by a woman who also must have been keenly aware of the depth and reach of her own remarkable nature.

Afterword

The Battle for the Brain Redux
Brain Imaging, Neurosexism, and Feminist Science

This book has demonstrated how Victorian women writers used the earliest theory of physiological psychology to make subversive arguments across a range of social domains. Some of these arguments were more overtly feminist than others, but the underlying implication that one's mental nature is often at odds with one's position in the world had a far more radical valence for women, who found in popular brain science confirmation of their unrecognized intellectual capabilities. For this reason, feminist phrenologists and women writers embraced an accessible form of biological science to challenge the limited and limiting categories of feminine identity created and maintained by both social construction and an emerging sexist strain of evolutionary biology. What we know (or think we know) about the brain, both in the nineteenth and twentieth century, has consistently tended to reflect, rather than create, cultural understandings of essential identity.

As with phrenology, contemporary brain science promises direct access to our innate psychology, but now that access is due to developments in noninvasive imaging technology including Positron Emission Tomography (PET), Single Photon Emission Computer Tomography (SPECT), and Functional Magnetic Resonance Imaging (fMRI). In the last decade, a number of global collaborative imaging projects have emerged, including ENIGMA (Enhancing Neuro Imaging Genetics Through Meta-Analysis), a network of 300 researchers across thirty-three countries analyzing the brain scans and genetic information of 30,000 people (Mohammadi 462) and the International Brain Initiative, which formed in 2018 to coordinate large-scale brain projects across the world (Bjaalie et al. 212). The most immediate potential benefit from such research is the treatment of neurological disorders, but it also has implications for contemporary understandings of the self. Scientists have begun to link localized brain activity to specific personality traits, including pessimism, risk aversion, persistence, empathy, and even spiritual belief (Olson 1548–49). As sociologist Nikolas Rose has observed, due to these new technologies, people's "personalities, capacities, passions, and the forces that mobilize them—their 'identities' themselves—appear to be explicable, potentially at least, in biological terms" (225).

Although phrenology and contemporary neuroscience are dissimilar in methodology, their use of technology, and their status in relation to the scientific establishment, they nevertheless share the same foundational premise and corollary: that the brain is the organ of mind, and that it is therefore theoretically possible to identify, chart, and observe the physiological signs of mental ability and potential behavior. Because the physiological basis of mind was first proposed in the nineteenth century, Victorian-era considerations of the philosophical, ethical, and social consequences of such knowledge acquire a new significance considering new advancements in neural imaging. Comparing the dissemination of recent neuroscientific research to nineteenth-century treatments of brain localization offers suggestive insights into the larger implications of recentering conceptions of human identity on a more rigorously biological basis. As a complete map of the neural circuitry of the human brain does not yet exist, exactly how this recentering will be culturally disseminated is still unknown. It is equally impossible to gauge whether the political, economic, and social applications of such information will ultimately prove felicitous, deleterious, or (what is most likely) a mixed bag. In many ways, where we are now in terms of penetrating the physiological nature of the mind is precisely where Gall believed he was in the 1790s.

As in the nineteenth century, the rush to confirm essential mental distinctions between the sexes has taken center stage in contemporary neuroscience. As neurobiological social philosopher Michael Gurian explains,

> Before these neural scans, we could say, "There's no such thing . . . as 'male nature' or 'female nature.' We all have the

same brain, which gets shaped into 'masculine' and 'feminine' by culture." [But] what we've learned from PET scans (as well as SPECT scans and MRIs), this sort of thinking is no longer valid. Though societies, cultural influence, and what is broadly called "nurture" do have much to do with the psychological costumes men and women wear, the male and female brain are "male" and "female" regardless of the culture . . . (6)

Working to counteract gender neutral parenting and education, Gurian has founded an institute that provides training for teachers and counselors. Psychologist Leonard Sax, another noted advocate for sex-specific approaches to education, has also vaunted the role of brain research in finally disproving the idea that differences in behavior between the sexes are caused by socialization rather than innate hardwiring. As he writes in *Why Gender Matters*, "gender is not primarily a social construct. It is a biological fact of our species, just as it is for gorillas and chimpanzees and every other primate" (290). The researcher on this subject whose work has received the most notoriety in recent years is Cambridge psychologist Simon Baron-Cohen, who developed the Sympathizing and Empathizing Quotient to statistically measure brain sex differences. He and his team claim that, on average, women tend to have a mind that naturally interests itself in the thoughts and feelings of others whereas the male brain is more given to building and understanding systems. Such hardwired differences, which Baron-Cohen speculates to be the outcome of tens of thousands of years of human evolution, align with demonstrated professional abilities and career preferences. Those with female brains excel in roles including "social workers, mediators, group facilitators, or personnel staff," whereas those with a male brain "make the most wonderful scientists, engineers, mechanics, technicians, musicians, [and] architects . . ." (185). Baron-Cohen is careful to clarify that one brain type is not necessarily superior to the other; they are simply different—a qualification similarly used to rationalize the separate spheres ideology of the nineteenth century (10).

Predictably, neuroscience has been used not only to explain but also to defend women's underrepresentation in scientific fields as an effect of innate biology rather than discriminatory practices, echoing the gambit of sexist scientists in the nineteenth century. While women were excluded from directly participating in their fields, Victorian evolutionary scientists and anthropologists sought to find an innate distinction in the brain that would prove men's intellectual superiority and thereby legitimate women's political, educational, and social subordination at a time when it was being questioned. Today, with STEM fields receiving increasing criticism for the disproportionate number of men in their ranks, scientists have turned to neuroscience to legitimate women's continued exclusion. Perhaps the most notorious instance of this was former Harvard President Lawrence Summers's speculation in 2005 that women's underrepresentation in STEM departments was tied to innate cognitive inequalities. At a conference convened specifically to address the issue of diversifying the workforce in science and engineering, Summers conceded the effects of socialization and discrimination but gave substantially more weight to "different availability of aptitude at the high end," both in terms of innate ability and intellectual taste. In the wake of the criticism that followed, cognitive psychologist and linguist Steven Pinker came to Summers's defense, claiming that the logic of evolutionary psychology would support a higher degree of variability in males because an exceptional male can sire more exceptional children in a lifetime than his female counterpart (16). He also outlined how damaging a continued belief in intellectual equality could be for women, potentially leading them to feel "pressured" into careers they may not like, and the possibility that installing gender quotas for grants might lead some to suspect that the (presumably few) deserving female grantees did not really merit their awards (15). As terrifying as it might be to find oneself an unfulfilled female physicist, Pinker found most disturbing the negative reaction to Summers's remarks: "Now anyone who so much as raises the question of innate sex differences is seen as 'not getting it' when it comes to equality between the sexes. The tragedy is that this mentality of taboo needlessly puts a laudable cause on a collision course with the findings of science . . ." (17). Yet, given the preceding decade's preoccupation with finding biological evidence of hardwired cognitive difference (including that which is cited in Pinker's own best-selling book), it is difficult to countenance this statement as accurate.[1] In fact, what was truly novel about the Summers incident was not his scientific claims but that an audience rejected them.

As the first chapter of this book argues, a key reason Victorian women gravitated toward phrenology in the first place was because it was one of the few sciences available to them due to the institutional barriers maintained by the scientific establishment. Thankfully, today women are in a far better position to directly participate in, and contest, brain science that explicitly or implicitly supports innate intellectual differences between the sexes. For instance, sociomedical scientist Rebecca M. Jordan-Young has discredited one of the most long-standing and uncritically accepted theories of sexual difference in mind, the organization–activation hypothesis. First proposed in 1959, this hypothesis holds that sexual differentiation in the brain is caused by its exposure to prenatal hormones that give rise to permanent masculine or feminine predispositions. An enormous body of brain organization research has since emerged in support, although the innatist premise of the hypothesis has remained largely unquestioned. Jordan-Young, however, analyzed every brain organization study from 1967 to 2000, as well as all major studies through 2008, and found that "the evidence simply does not support the theory" (xiii). As she explains, these studies are limited from the start because they must rely entirely on quasi-experimental evidence, as it is obviously unethical to conduct controlled experiments on the effect of prenatal hormones in humans. Without true experiments, causality can only be empirically justified through the consistent findings across many studies. Yet, when the actual data on brain organization research is taken together, the conclusions are inconsistent and contradictory (42–54). In addition, she found that review articles and literature reviews "systematically omitted" data that demonstrated the ease with which behavior could be modified by environmental factors (286).

The failure of scientists to recognize the brain's plasticity when accounting for observed mental differences between the sexes has also led to slippage between the physical and the innate, despite the fact that the brain records and encodes the effects of all kinds of experiences in neural circuitry, including gendered differences in socialization. Neuroscientist Lise Eliot claims that this "notion that sex differences in the brain, because they are biological, are necessarily innate or fixed is perhaps the most insidious of the many public misunderstandings on this topic" (897). Sociologist Cordelia Fire attributes this tendency in part to the visual nature of brain scans, which makes studies appear substantive when they actually just re-describe experiments' results without providing an explanation for the cause of the observed distinctions. While such claims might appear to have demonstrated cause because something that was not visible before now appears as a glowing region of the brain, it is far more likely that observed differences in scans show the effects of highly gendered environment (170–71).

Beyond critiquing the faulty logic that underlies "neurosexism," female scientists have also started to amass a formidable body of evidence that demonstrates the lack of intrinsic cognitive difference between the brains of women and men.[2] Daphana Joel and her team at the Tel Aviv University in Israel scanned the brains of 1,400 people between the ages of thirteen and eighty-five, and found that most brains are highly individualized "mosaics" of characteristics traditionally recognized as male or female. This not only challenges the idea of sex in mind but also lends scientific support to what many theorists have long suspected: that gender is not binary. In a major blow to defenders of unequal gender representation in STEM professions, Jessica Cantlon and Alyssa Kersey at Carnegie Mellon used fMRIs to observe the neural processing of mathematics in 104 children between the ages of three, which showed indistinguishable patterns of brain activity. This tracks with Harvard psychologist Elizabeth Spelke's 2005 finding that across 111 studies on gender differences in math and science, men and women on the whole demonstrated equal levels of aptitude and performance, even as early as six months of age. In the same year, psychologist Janet Hyde analyzed forty-six meta-analyses of sex difference studies and found that males and females were found to be similar across most psychological variables, leading her to formally propose the "Gender Similarities Hypothesis" to replace the "differences model" that has long dominated psychology (581). As this brief but representative survey shows, the scientists who have challenged sexual dimorphism in the brain have overwhelmingly been women, who, much like the nineteenth-century women interested in brain science, came to completely different conclusions than their male counterparts. Today, however, women conduct their research as distinguished professors at top-tier research universities using state-of-the-art technology. Although prejudice and discrimination still prevent many women from having successful careers in science, today women can become educated and credentialed in any field, and need not find alternate routes to their interests through more accessible, but less respected, popular sciences.

In the hands of feminist neuroscientists, a discipline-specific essentialism is now being used to show the absence of sexual distinction in the mind's intellectual capabilities, but this very essentialism also highlights the degree to which social construction contours our psychology and literally encodes itself into our neural pathways. In making this known and clear, practices that instill and reify inequitable treatment on the basis of gender can be interrupted, removed, and replaced. As neuroscientist Gina Rippon puts it, "a gendered world will produce a gendered brain . . . understanding how this happens and what it means for brains and their owners is important, not just for women and girls, but for men and boys, parents and teachers, businesses and universities, and for society as a whole" (xxi). The implications of an essentially equal but environmentally malleable brain were similarly identified by the Victorian women discussed in this book, and the value in knowing this lies less in appreciating their prescience than becoming aware of the necessity for continued vigilance. Essentialist claims about the nature of mind have continually underwritten significant reconsiderations of major ideological structures, particularly where sex is concerned. The fact that we actually still know so little about the brain has not curtailed the stories we tell about it, and these stories clearly have the power to reinforce or overturn popular conceptions of individual and collective identity. We might do well to revisit nineteenth-century considerations of the social, ethical, and philosophical implications of a truly legible mind because in the very near future it will be within our grasp.

Notes

Introduction

1. Astronomer and scientific polymath John Hershel was the president of the Royal Astronomical Society and author of the influential *Preliminary Discourse on the Study of Natural Philosophy*, which established the importance of inductive reasoning in the pursuit of scientific knowledge. Although he received no formal training, George Henry Lewes's investigations into animal physiology and physiological psychology solidified his position within the scientific establishment. For an account of Lewes's contributions to the early development of physiological psychology, see chapter 7 of Rick Rylance's *Victorian Psychology and British Culture*.

2. See Bordo's *Unbearable Weight* and Spillers's "Mama's Baby, Papa's Maybe."

3. Rosalyn Diprose identifies ideological discourse about embodied sexual difference as the prime source of women's subjugation. As she puts it, "the moral, legal, industrial and interpersonal evaluation of sexual difference is productive: it produces the modes of sexed embodiment it regulates [and] any injustice experienced by women begins from this mode of production and maintenance of sexual difference" (viii). While Diprose locates sexual difference as the central impediment to women's liberation, she sees no way "outside" this differential system since "we are the embodied products of regimes which regulate sexual difference [and] these regimes support our existence. Thus, there is no back door to freedom . . ." (131).

4. Thomas Laqueur describes this as a movement from a "one sex" to a "two-sex" model, in which a "biology of cosmic hierarchy gave way to a biology of incommensurability, anchored in the body" (207).

5. Schiebinger points out that in early modern Europe women were importantly involved in scientific circles but that complementarity was the death knell of women's publicly accepted (or tolerated) involvement in science (*The Mind has No Sex?* 230–37).

6. In addition to these accounts by historians of science, Patricia Murphy traces the influence of sexist scientific discourse on literary works in *In Science's*

Shadow: Literary Constructions of Late Victorian Women. She claims the increased scientific interest in finding evidence of women's biological inferiority can be directly attributed to agitation over the Woman Question (1–2).

7. See, for instance, Shapin's "Phrenological Knowledge and the Social Structure of Early Nineteenth-Century Edinburgh," Winter's "Construction of Orthodoxies and Heterodoxies in the Early Victorian Life Sciences," and Wallis's important collection, *On the Margins of Science: The Social Construction of Rejected Knowledge*.

8. Separate and distinct scientific disciplines began to emerge as such only in the nineteenth century, when modern science underwent a process of professionalization stretching roughly from the 1830s to the 1860s.

9. For an account of the relationship between professionalization and women's marginalization in botany, see chapter 6 of Ann B. Shteir's *Cultivating Women, Cultivating Science*.

10. Ann B. Shteir argues that the emergence of women's popular science writing in the Victorian period was a consequence of increasing professionalization, which effectively relegated women to the more marginal "arenas of 'amateur' and 'popular' science" ("Elegant Recreations" 236).

11. Alison Winter identifies a number of professional women mesmerists, including one permanently employed at the Dublin Mesmeric Infirmary run by the archbishop of Dublin, Richard Whately (*Mesmerized* 138, 156). In *The Darkened Room*, Alex Owen argues that women played a central role in the spiritualist movement as both private and professional mediums, and further, that their practice often subverted conventional notions of Victorian femininity (see especially chapter 8, "Spiritualism and the Subversion of Femininity").

12. In *Sexual Science*, Russett argues that this concerted effort to scientifically demonstrate women's inferiority was in direct response to feminist agitation, and that it involved the fields of "Anatomy and physiology, evolutionary biology, physical anthropology, psychology, and sociology," all of which "were guided, with few exceptions, by the beacon of evolution . . ." (10). For a detailed account of the connections between evolutionary theory and the argument for the complementarity of the sexes, see especially chapters 2–5.

13. Dames defines this physiological theory of the novel as follows: "the novel of the nineteenth century trained a reader able to consume texts at an ever faster rate, with a rhythmic alternation of heightened attention and distracted inattention locking onto ever smaller units of comprehension" (7).

14. Ilana Kurshan's examination of early Victorian phrenology journals demonstrates how phrenologists used literary techniques and quotations to demonstrate its authenticity and promote its authority, but she is not concerned with how literary writers used phrenology.

15. Notable books on evolutionary theory and Victorian literature include Gillian Beer's *Darwin's Plots*, George Levine's *Darwin and the Novelists*, Joseph Carroll's *Literary Darwinism*, and more recently Alexis Harley's *Autobiologies*. As

Jay Clayton observes, "Most books on literature and science in the Victorian period do not even begin until the 1860s, and thus inevitably, focus on the Darwin controversies" (*Dickens in Cyberspace* 223 n.11).

16. Phrenology was one of the first phenomena to be salvaged for serious consideration by historians of science. For an account of how social agendas were transferred to debates about phrenology in Edinburgh, see Steven Shapin, "Phrenological Knowledge and the Social Structure of Early Nineteenth-Century Edinburgh." For an account of how phrenology functioned in legitimating the interests of the upwardly mobile middle class in particular, and ideological interests of industrial capitalism in general, see Roger Cooter's study of phrenology's influence in Britain, *The Cultural Meaning of Popular Science: Phrenology and the Organization of Consent in Nineteenth-century Britain*. For a sociological account of how phrenology served to both reinforce and challenge social, political, and disciplinary boundaries, see Thomas F. Gieryn, *Cultural Boundaries of Science: Credibility on the Line*, chap. 3.

17. *The Constitution of Man* sold 100,000 copies between 1828 and 1860, whereas Darwin's *Origin of Species* sold 50,000 copies between 1859 and 1900 (Cooter, *Cultural Meaning* 120).

18. For a nearly complete catalog of the known bibliographic sources related to phrenology in Britain in the nineteenth century (listing 2,207 references and 56 phrenological societies), see Roger Cooter, *Phrenology in the British Isles: An Annotated, Historical Bibliography and Index*.

19. See, respectively, Clarke, Warne, Lyons, and Browne.

20. A longstanding critic of phrenology, Lewes tried to separate Gall's physiological approach from the phrenologists who followed in his wake, valuing the former and dismissing the latter.

21. In addition to Shuttleworth's book, some of the more notable studies of Charlotte Brontë's use of phrenology include those by Nicholas Dames, Alan Rauch, Mary Armstrong, Nathan Elliot, and Leila S. May.

Chapter 1

1. Broca based his claim about the increase in the difference of size over time on the prehistoric skulls of only seven males and six females (Gould 105). For his modern comparative study, Broca selected older women and taller than average men, two attributes which have an influence on brain size. In reanalyzing the data to take into account the relative mass when controlling for height and weight, Gould concludes the "true figure [of difference] is probably close to zero and may as well favor women as men" (106).

2. As Evelleen Richards points out, Vogt's "woman-as-child-as-primitive argument . . . provided the sole scientific underpinning of Darwin's conclusions on the futility of higher education for women" (74).

3. Evelleen Richards makes this connection between Mill's presentation in the House and the sudden spike in publications on women's inferiority ("Huxley" 269).

4. Sharples gave a series of phrenology lectures through Carlisle, London, and Gravesend in the 1830s, and also delivered a lecture on phrenology at the London Hall of Science in 1839 (Cooter, *Phrenology* 49)

5. Written by Emily G.S. Saunders, Annie I. Oppenheim, Ida Mitchell Ellis, and Jessie A. Fowler, respectively.

6. As Roger Cooter notes, women were invited to attend special meetings of phrenological societies "partly as a radical gesture, partly for bumping up attendance figures" (*Cultural Meaning* 369 n.89).

7. Spurzheim himself claimed that when it came to intellectual faculties, "men undoubtedly enjoy the superiority" while women are naturally "subordinate to men" (227, 229). Another phrenologist claims that the psychological development of women's brains reveals that "Women may strive in vain; she will never, except in a few cases, depart from the sphere in which her native destiny has placed her" (Guillot 194). Such statements, however, are actually quite rare—phrenological works by men tended not to dwell on the mental differences between the sexes.

8. The report for the meeting lists twenty-two members by name, twelve of whom were women (see "British Phrenological Association" in *Phrenological Magazine* 5 n.s. [1889]: 182–97, 182). Two of them had previously been appointed to the council, the professional phrenologist Annie I. Oppenheim and a Miss Baker (see "British Phrenological Association" 4 n.s. (1888): 167–72, 168).

9. Patenall's first address to the Hasting's branch was published in *The Phrenological Magazine* ("Paper Read"). She provides her credentials in a letter to the editor of the *St. Leonard's Observer*, republished in the *Magazine*, defending her standing as a representative of the British Phrenological Association in the area ("To the Editor").

10. For an example of a warmly received paper delivered by Annie I. Oppenheim, see "British Phrenological Association" in *The Phrenological Magazine* 5 n.s. (1889): 64–68. *The Phrenological Magazine* was incorporated into the *Phrenological Journal and Science of Health* in 1897. The articles published by women are too many to list here, but some representative articles include Mrs. Henry Wallenstein's "Child Study and Phrenology," Margaret E. Parker's "Healthy Work for Women," Annie N. Patenall's "Character," and Susanna W. Dodds's "A Healing Art in the Twentieth Century."

11. While the Fowler family was American, their phrenological empire in England was considerable, from Lorenzo Niles's and Lydia Fowler's emigrations to England in 1860 until Jessie Fowler's return to America in 1896. I have limited my exploration of the Fowlers to the British context. For a treatment of the Fowlers' activities on both sides of the Atlantic, see M.B. Stern's *Phrenological Fowlers*.

12. Lydia Becker delivered five lectures to the BAAS between 1868 and 1875 (Bernstein 90). She also wrote on both botany and astronomy and corresponded with Charles Darwin (86).

13. This paper was published in *The Englishwoman's Review* under the title "Is There Any Specific Distinction Between Male and Female Intellect?" All references will be to this article.

14. The increased inclusion of phrenological biographies of women under Jessie Fowler's editorship contrasts with the practice of Alfred T. Story, *The Phrenological Magazine*'s former editor, who opened each issue with a portrait and phrenological illustration of illustrious men, such as John Ruskin, Cardinal Newman, and Alfred Tennyson.

15. Practical phrenologists varied in the number of manifestations organs could exhibit, but I have never seen a delineation booklet with less than four. Phrenologists also took into account the client's temperament (lymphatic, sanguine, bilious, or nervous). Thus, the least number of possible personalities in the phrenological schema would be 4×4^{36} or 4.7×10^{21}.

16. The novel is serialized in *The Phrenological Journal and Science of Health* from 1897–1898.

17. This segment of the novel appears in *The Phrenological Journal and Science of Health* vol. 104, no. 2 (1897): 230–33.

18. See also "Women Physicians in the United Kingdom," "New Hospital for Women," and Jessie Fowler's "Women's Work in Art."

19. See "Putting Character First: The Narrative Construction of Innate Identity in Phrenological Texts."

20. The author of the article on Fairman acknowledges prevalence of such profiles, calling them "Phrenological Sketches of Women Engaged in Medicine, Philosophy, Teaching, or Business" by way of adding "Law" to the list (45).

Chapter 2

1. As of 21 September 2017, using a subject search for "phrenology," the MLA Bibliography lists critical studies on only three (English) Victorian literary figures: George Eliot (5), Charles Dickens (3), and Charlotte Brontë (11). The only other writer whose use of phrenology has been studied to a similar extent is Edgar Allan Poe (eight critical works). Nicholas Dames says that in Charlotte Brontë's work "phrenology achieves its greatest visibility in nineteenth-century fiction" ("Clinical Novel" 369).

2. Nicholas Dames's reading initially appears to differ with Shuttleworth's insofar as he stresses that Brontë embraces the science for its potential to make character immediately legible via visual surfaces, as opposed to Shuttleworth's claim that phrenology gave Brontë a theory through which she could articulate psychological depth. Both Dames and Shuttleworth, however, interpret Brontë's

ideological use of phrenology through a Foucauldian lens, emphasizing the ways in which she used the science to embrace the value of self-monitoring, self-discipline, and surveillance. Along similar lines, Rauch argues that in *The Professor*, Brontë uses phrenology as a discourse of social order that allows Crimsworth to improve his position in the world both economically and biologically through his knowledge of science. Armstrong's reading examines how phrenology helps to produce the "multiple erotic strands" of *Jane Eyre* (107). In opposition to the majority of critics who have viewed Brontë's perspective on phrenology as largely uncritical, Elliot and May have argued that by the time she got to *Villette*, she had grown suspicious of the science and explored its negative implications in her final novel.

3. In fact, in explaining her decision to focus exclusively on Charlotte in her study, Shuttleworth acknowledges that among the sisters Anne was the one who "foregrounded" phrenology in her work, whereas Emily and Charlotte referenced the science more subtly by representing psychological energies and exchanges (5).

4. As Tamara S. Wagner comments in reference to Shuttleworth's *Charlotte Brontë and Victorian Psychology*, "no comparable study has so far been attempted to reassess the other Brontë sisters' more muted interest in clinical discourses" ("Speculations" 137 n.25). Although in the case of phrenology in *Tenant*, Anne Brontë's interest is certainly not "muted," the observation stands: outside of passing references, the only work to address the science in a sustained way being Marianne Thormählen's "The Villain of *Wildfell Hall*." Thormählen's essay is less a reading than a useful contextualization of the novel's depiction of Arthur Huntingdon in relation to contemporary texts on phrenology and alcohol addiction.

5. For a discussion of the promotion of phrenology within mechanics' institutes, see chapter 7 of Roger Cooter's *Cultural Meaning of Popular Science*, particularly pp. 145–51.

6. The Brontës borrowed *Blackwood's* from a neighbor and subscribed to Fraser's starting in 1832 (McDonagh xxxii). For information on *Blackwood's* virulent opposition to phrenology, see Strachan.

7. Articles on the materialist implications of phrenology in *Fraser's* appearing after 1832 (when the Brontës subscribed) include "Mr. George Combe and the Philosophy of Phrenology" and "Mares'-nests Found by the Materialists, the Owenites, and the Craniologists."

8. Harriet Downing's "Remembrances of a Monthly Nurse" were published in both *Fraser's* and *The Monthly Magazine*, and after her death the tales were collected and published as a book by the same name. The advertisement for the volume notes that the tales "attracted considerable attention at the time" and that many originally believed them to be authored by William Maginn (C.T.D.).

For more on the popularity of the tales, see *The Metropolitan Magazine's* "A Few Words Upon the Life and Writings of the Late Mrs. Harriet Downing."

9. Shuttleworth discusses Patrick Brontë's references to *The Philosophy of Sleep* in his annotations but does not note Macnish's use of phrenology in this or other works.

10. The preface to the second edition (1834), cited here, is essentially a defense of the validity of phrenology. Macnish also employs phrenological discourse in the first edition, but expands its use in the second. The 1841 catalog record for the Keighley Mechanics' Institute (published in "Where the Brontë's Borrowed Books") does not specify which edition was in the collection. Macnish was also the author of the *Introduction to Phrenology* (1836) and *The Anatomy of Drunkenness* (1827).

11. Shuttleworth makes note of this experience (*Charlotte Brontë* 57 and 259 n.2), as does Nicholas Dames ("Clinical Novel" 367) and Ian Jack (390–91).

12. In a letter to Elizabeth Gaskell, Mary Taylor remarked that her friend Charlotte was "an ugly woman" with a "square face and large disproportionate nose" (qtd. in Barker 761). In her biography of the Brontës, Juliet Barker observes that Charlotte Brontë was well aware of her lack of beauty and that she remained "hypersensitive about her appearance" throughout her life (760).

13. This chalk portrait, made by George Richmond, was reproduced in Elizabeth Gaskell's *The Life of Charlotte Brontë*. Juliet Barker describes the sittings as "nerve-racking" for Charlotte, who was reluctant to have "a portrait made of what, she was all too well aware, were neither beautiful nor attractive features" (761, 760). She first burst into tears when asked to remove an unconvincing hairpiece that the artist mistook for a curious bit of fabric on her head, and cried again after seeing that the flattering portrait bore less resemblance to her than to her more attractive sister, Anne (Barker 761).

14. Roger Cooter provides a sociopolitical analysis of the opponents and proponents of phrenology in Britain prior to 1835. Whereas most antiphrenologists were members of the upper class and politically conservative, the proponents of the science were young, politically liberal, and of comparatively lower economic status (42–48). He concludes that on the whole, opponents of the science were "anxious to preserve institutionalized social and ideological interests" (*Cultural Meaning* 47).

15. Nancy Stephan discusses the connection between phrenology and the development of biological racism in chapter 2 of *The Idea of Race in Science*.

16. The significance of Bertha's race is most famously invoked in Gayatri Spivak's reading of the novel, in which she identifies Bertha as a "white Jamaican Creole," which Brontë positions somewhere between animal and human (247). Spivak argues that the novel's "feminist individualism" emerges in relationship to Bertha's identity as abject other, who is, unlike Jane, "excluded from any share in

this emerging norm" ("Three Women's Texts" 244–45). Susan Meyer claims that the novel most notably constructs Bertha as black, and sympathetically aligns her with the "desire for revenge on the part of colonized peoples" (69). More recently, Patricia McKee has argued that Brontë uses the ambiguity of race to strategically identify "not only the Creole Bertha Rochester but the upper-class Blanche Ingram to the racial status of dark primitives" (68).

17. This is a point phrenologists often made about "inferior" races. For example, as Combe says in *The Constitution of Man*, "It appears to me that the native American savages, and native New Hollanders, cannot, with their present brains, adopt Christianity or civilization" (167). Combe held that all races could improve the quality of their cerebral development through a knowledge of organic law and selective breeding, but he framed this as a very gradual process (166–67).

18. Combe notes in the *Constitution* that "a small development of the organ" of benevolence leads to an "indifference to the welfare of others" (*Elements* 93).

19. A reviewer for *Sharpe's London Magazine* was confused about the sex of the author of the novel, reasoning, "none but a man could have known so intimately each vile, dark fold of . . . an exhibition as this book presents," yet speculated only a woman "would have written a work in which all the women . . . are so far superior in every quality, moral and intellectual, to all the men . . ." (Review of *The Tenant*, 184). Although he questioned the book's aesthetic merit, Charles Kingsley defended its realistic depiction of debauchery, observing that "a certain degree of coarse-naturedness . . . may be necessary for all reformers, in order to enable them to look steadily and continuously at the very evils they are removing" (424). For recent criticism on the novel's progressive gender politics see Berry, Jacobs, Joshi, and Morse.

20. For a rhetorical analysis of delineation booklets and similar species of phrenological paraphernalia, see my article "Putting Character First: The Narrative Construction of Innate Identity in Phrenological Texts."

21. The words "predominate" and "preponderate" were used (to exhaustion) by phrenologists to describe the relationships between the three regions of the brain: intellectual, moral, and animal. For example, Combe comments that Melancthon's skull shows "the decided predominance of the moral and intellectual regions over that of the animal propensities," whereas in the head of William Hare "the animal propensities decidedly preponderate over those of the moral sentiments and intellect" (*Constitution* 138, 137).

22. The exchanges over the sketches occur in chapter 18, culminating in Helen burning his complete, colored miniature portrait as a way of disproving her (actual) desire to keep it (156).

23. Shuttleworth reads this scene as "One of the clearest statements of the phrenological doctrine of self-improvement," which it undoubtedly is (*Charlotte

Brontë 67). What I hope to point out, however, is that this side of phrenological discourse is undercut in the novel by its more deterministic aspects that stress natural limits and the ultimate failure of improvement for extreme cases.

24. This split is discussed at greater length in chapter 3.

25. This felicitous conclusion that secures a better Arthur for the future through biology is perfectly in line with phrenological discourse that stressed the primacy of the mother's development over the father's in determining the child's mental capabilities. As Combe puts it, "The character of the mother seems to have the chief influence in determining the qualities of the children, particularly where she has much force of character, and is superior in mental energy to her husband . . ." (*Constitution* 148).

26. As Spurzheim puts it, "The innate constitution . . . which depends on both parents . . . is the basis of all future development" (*Education* 59).

27. See, for instance, "South Sea Islanders," "On the Development and Character of the North American Indians," and James Montgomery's "An Essay on the Phrenology of the Hindoos and Negroes."

28. Margaret Shaw makes a similar point about Lucy in *Villette*: "Lucy turns others into types and emblems . . . her use of the foreign (of Villette and Catholicism), for example, comes close to territorializing—a way of consolidating the self . . ." (826).

29. Simpson adapts this table from a similar one presented by "A Mother" in an article published by the *Phrenological Journal*. The author details how she used the record book on her own children to great effect ("Phrenology Applied").

30. For instance, Gwen Hyman argues that Arthur's drinking and consequent bodily degeneration displays how the aristocratic man is constitutionally unfit for the new Victorian culture of productivity and self-management. Beth Torgerson reads Arthur's alcoholism as a symptom of his excessive masculinity (because he does not moderate) that Helen's excessive femininity (because it is too self-sacrificing) fails to regulate. Torgerson concludes that Helen's failure "encourages readers to question an outdated cultural script for such extreme forms of femininity and masculinity" (35). I agree that Brontë critiques this domestic ideal by displaying its inefficacy, which the novel makes quite clear when Helen repeated and emphatically absolves herself from any blame for Arthur's decline. I argue, however, that Arthur's failure to reform is not due to his over-investment in any cultural ideal but to his biological inability to change, which is supported by the novel's sustained interest in scientific discourses that emphasize the role of essentialism in determining character.

31. Caldwell also notes that in postmortem examinations of the brains of drunkards "the animal compartment of the brain has been found preternaturally large, in proportion to the others . . ." (341).

32. Caldwell also points to drunkenness as proof of hereditary transmission (341–42).

33. Robert Macnish cites Caldwell's phrenological diagnosis of addiction and recommended use of tartar emetic in *The Anatomy of Drunkenness*, adding that his essay is "one of the ablest papers which has hitherto appeared upon the subject" (157).

Chapter 3

1. Martineau revoked this request in the 1872 codicil to her will, although she explains in no uncertain terms that this change had nothing to do with her scientific allegiances: "I wish to leave it on record that this alteration in my testamentary directions is not caused by any change of opinion as to the importance of scientific observation on such subjects but is made in consequence merely of a change of circumstances in my individual case." According to an 1884 biography by Mrs. F. Fenwick Miller, Martineau revoked the earlier request because Atkinson was residing outside of England and was therefore "not in a position to usefully accept the bequest" (209). This, however, does not appear to be the real reason for his refusal since Martineau stipulates in the will that her brain should be delivered to him only "if [her] death should take place within such distance of the said Henry George Atkinson's then present abode." Further, Martineau's skull would not be affected by decay. Atkinson's concern about the organic integrity of her brain after death is probably a polite excuse made to cover his reluctance to accept her well-meant, if undesired, bequest.

2. Initially, Martineau planned to donate her body to science due to the shortage of bodies for dissection. After the passage of the Anatomy Act in 1832, however, she decided her particular donation would no longer be necessary, and she changed her will. Later, however, she learned (probably from Atkinson) that "easy as it is to procure brains and skulls, it is not easy to obtain those of persons whose minds are well known, so that it is rather a rare thing to be able to compare manifestations with structure . . ." (*Autobiography* 1: 391).

3. Hereafter referred to as *Letters*.

4. Diana Postlethwaite deems the publication "close to the lunatic fringe," and only useful as "a fascinating case study of the way in which lesser Victorian minds seized upon seminal ideas of the period [mesmerism and phrenology]" (*Making It Whole* 142). Valerie Pichanick comments that the work is "not intrinsically important. It does not deserve a special place in the hierarchy of Victorian literature or philosophy," locating its value in "what it tells us . . . about Harriet Martineau" (187). Shelagh Hunter similarly comments that "The *Letters* are significant in Martineau's emerging autobiographical story, not for their overt promotion of a short-lived and controversial science . . ." (102).

5. Martineau argues for women's equal opportunity to gain employment at every level, including the learned professions, in "Female Industry" and claims

women should receive equal pay for equal work in an article for the *Daily News* (2 April 1856). She wrote in support of women's equal access to education throughout her career but perhaps most forcefully in "On Female Education" and "What Women Are Educated For." In *Society in America*, Martineau claims that the core principle of the Declaration of Independence—that just governments operate at the consent of the people—is completely incompatible with the status of women and enslaved people, who have never consented to their political nonexistence. For an account of Martineau's involvement in the repeal of the Contagious Diseases Acts and how it aligns her with first-wave feminism, see Logan (159–63).

6. See Webb (303–9), Logan (200), and Pichanick (197–98).

7. Caroline Roberts's chapter on the *Letters* in *The Woman and the Hour: Harriet Martineau and Victorian Ideologies* is one of very few accounts of Martineau's use of phrenology that neither dismisses nor makes excuses for her interest in the science. Roberts also views her use of the science as subversive, noting that the system of phrenology outlined by Atkinson and Martineau departs in significant ways from the more hierarchical arrangement of faculties in works by Combe and Spurzheim. Thus, whereas Combe's articulation of phrenology was complicit in naturalizing a bourgeois ideology, Martineau and Atkinson's version was more radical due to its egalitarian implications (181–86).

8. By at least 1855, Martineau was aware of an extremely large abdominal tumor, which she describes in a letter to John Chapman (*Selected Letters* 130). The symptoms Martineau experienced were most likely caused by the pressure exerted by the cyst on her heart and lungs. Posthumous medical accounts of Martineau's illness claim that the mesmeric therapy happened to coincide with the tumor's expansion from the small pelvic cavity into the larger abdominal region, resulting in a temporary release from internal pressure (Ryall 40, 45). Although her symptoms returned a decade later, in her *Autobiography* Martineau still defended mesmerism against the professional intolerance of medical men (2: 194).

9. Subsequent references to Atkinson will be from this work.

10. Atkinson explains, "I make mesmerized patients hold a weight, and tell me where it influences the brain, and see how the excitement of that part affects them, and I cause them to trace their various sensations, injuries, &c., to the brain" (64). He further remarks that the "knowledge of these organs of the cerebellum" is "highly important towards the explanation and cure of disease; and to be most suggestive and necessary to those who mesmerize. Every mesmerizer should understand phrenology and phreno-mesmerism" (65).

11. Reprinted as *Letters on Mesmerism* (1845).

12. On Martineau's mesmeric cure and its connection to scientific professionalization in the medical community, see Ryall. For a discussion of Martineau's reclamation of femininity through her mesmeric account, see Postlethwaite's

"Mothering and Mesmerism in the Life of Harriet Martineau." On the connection between mesmerism and Martineau's continued interest in economic theory, see Ketabgian.

13. The "Report of the Proceedings of the Phrenological Association" appearing in the *Phrenological Journal* offers a condensed version of the address with a record of comments and notes. The complete address originally appeared in *The Medical Times* in 1842.

14. While not recorded as voicing an opinion on Engledue's address, Henry George Atkinson attended the annual meeting as one of the twenty-four gentlemen acting as the administrative committee of the Phrenological Association, and also delivered a paper titled "On Mesmero-Phrenology" on the Thursday session of the meeting, 23 June 1842.

15. For a clear and concise survey of the book's content, see Hoecker-Drysdale, pp. 155–58.

16. In reply to the long and boldly anti-theistic letter in which Atkinson equates scientific prophets with Christ, Martineau writes, "I am glad I asked you in what sense you used the words 'God,' 'Origin,' &c., for your reply comes to me like a piece of refreshing sympathy,—as rare as it is refreshing" (*Letters* 216).

17. Abrams discusses all but the last of these qualities of conversion in his analysis of Augustine's *Confessions* (84–85).

18. Hoecker-Drysdale solves this contradiction by claiming that Martineau attempts to "protect herself by focusing on Atkinson's views" (160), adopting a sycophantic position because her "literary success and economic independence were insufficient, it would appear, to convince her that she could 'go it alone' in presenting the challenges of science to conventional thinking" (158). Martineau's estimation of how the public would react must have been inordinately short-sighted if this were true because as the more famous co-author she bore the brunt of the criticism.

19. A treatise, autobiography, or apologia does not require an addressee beyond the reader, but the public confession must include a third position.

20. Along with *Pilgrim's Progress* and *The Constitution of Man*, Martineau cites *Robinson Crusoe* and the Bible as the four books reliably found on any bookshelf in Great Britain in the nineteenth century (*Biographical Sketches* 265).

21. See Abrams 47–48.

22. Martineau writes that the "argument of Compensation, by means of a future life, appears to me as puerile and unphilosophical as the Design argument in regard to 'Creation,' or the existence of things . . ." (164).

23. As Abrams notes, Augustine "adumbrates the concept of what we now call unconscious motivation, by internalizing the distinction, common in the Christian view of history, between the secondary causes which are available to human observation and the omnipresent but invisible First Cause" (85).

24. Martineau writes: "we are free to recognize things as they are,—to number our faculties, and to enlarge the number, if occasion should arise, by the development of new,—that is, hitherto unknown—faculties, which bear a relation to external things" (119).

25. John Chapman accepted the manuscript after Moxon's rejection.

26. In a letter to Helen Martineau on 14 July 1851, Martineau similarly observes that "People do not answer reviews" (*Collected Letters* 3: 204). On 1 September 1872, Helen Brown wrote to Arthur Arnold about the negative review of the *Letters* in the *Echo*, observing, "It has been her practice through life to leave unnoticed such mistakes in regard to matter which is before the public."

27. Although the unnamed woman eventually sought reconciliation, Martineau felt, as she had toward James, that "she had done too much," thus rendering their relationship unsalvageable (*Autobiography* 2: 353).

28. For instance, "The Pope versus Phrenology" prints in full an 1837 Papal decree denouncing a series of books that included a French translation of one of George Combe's phrenology manuals, and "Recent Attacks on Phrenology" identifies as *"foes"* those periodicals that print articles critical of the science (262).

29. Martineau's correspondence with Patrick Brontë and Arthur Nicholls concerned what she viewed as Charlotte Brontë's unflattering depiction of her (Martineau's) character as it appeared in Gaskell's *The Life of Charlotte Brontë*.

Chapter 4

1. In his "Afterword" to the Modern Library edition of the novel, Chis Willis identifies *The Trail of the Serpent* as "probably the first British detective novel" (408), qualifying his statement with the observation that while Charles Dickens's earlier *Bleak House* (1853) includes a police detective plot, that plot is not the central concern of the novel.

2. As Andrew Mangham puts it, the text has generated "a surge of interest in recent years" ("'Drink it up dear'" 95). For critical treatments that take a disability studies approach, see Ferguson and Tomaiuolo. Issues of Victorian concern addressed by scholarship include infanticide (Mangham, "'Murdered at the Breast'"), the commodification of children (Wagner, "'We have orphans'"), toxicology (Mangham, "'Drink it up dear"), and modern visuality (Green).

3. For explorations of the novel's relation to genre, see Bennett, Ferguson, Waters, and Willis.

4. The "Cases and Facts" section was included in almost every number of the "New Series" (1838–1847) of the *Phrenological Journal and Miscellany*. It was reserved for correspondents who provided evidence of phrenology's validity, either in the form of a "Fact" (demonstrating the correspondence between

skull measurements and known character), or in the form of the "Case," which presented a behavioral mystery solved through a phrenological reading. The majority of these cases fall into one of two pathologies—the psychological complaint, or criminal behavior.

5. The London phrenologist James Deville makes this statement by way of explaining why the majority of the skull casts in his extensive Phrenological Museum were of criminals. Deville's collection of 2,400 items included casts of living persons, animals, busts of eminent men, and skulls from different nations. Among the casts of "social types" Deville includes in his extensive collection, the largest section contained the heads of 150 "criminals, English and foreign" (Deville, "Account" 19). A close second were the 120 "pathological cases illustrative of insanity." The casts of other social types is as follows: writers (80), pugilists (25), musicians (70), persons devoted to religion (50), artists (40), and navigators (30). A discussion on Deville's collection and its connection to social typing appears in chapter 5. The catalog for Frederick Bridges's Phrenological Gallery in Liverpool lists 15 criminal skulls or casts (out of a collection of 96), including the usual suspects of William Burke (Bridges 10), David Haggart (45), John Thurtell (46), and Mary McInnes (9).

6. To list only a few: John Elliotson's "An Account of the Head of Rush, the Norfolk Murderer" (1849), "Case of William Saville" (1844), W.R. Lowe's "Case of John Williams" (1843), "Homicidal Monomania and Murder by Premeditation [Case of Mattos Lobo]" (183). James Bloomfield Rush was a tenant farmer who murdered the estate owner, Isaac Jermy, and his son in 1849. William Saville murdered his wife, Anne, and their three children with a razor in 1844. John Williams murdered the shopkeeper Emma Evans for her purse. Mattos Lobo was a Portuguese student who murdered his aunt, Donna da Costa, along with her son, daughter, and servant in 1842. The publication of such accounts was so frequent that Lowe begins his own case with the justification, "although a sufficient number of murderers' developments has been already published, to convince every one not willfuly [sic] blind, that their heads differ materially in shape from those of virtuous and superior persons,—yet the phrenologist cannot be provided with too ample an array of well authenticated facts . . ." (62).

7. For other case studies with a similar investigative bent, see "Case of [Francis Benjamin] Courvoisier, lately executed for the Murder of his Master, Lord William Russell" (328–29) and Otto, "Remarkable Case of a Sudden Morbid Excitement of Destructiveness."

8. The arc of this narrative structure resembles that of psychoanalytic recovery in terms of bringing to light a formative aspect of self that was always there but remains unknown. While phrenology posited and popularized the existence of a hidden self, its assumptions about the mind were fundamentally opposed to the Freudian concept of the unconscious. The nascent identity phre-

nology posited, while instinctual and highly influential, was not contoured by repression. The phrenological encounter did not aim at uncovering an unconscious contoured by personal experience but, rather, a physiological identity created by nature prior to experience.

9. The writer, C.J.A. Mittermaier, was a professor of criminal law at the University of Heidelberg who corresponded with Combe in 1842 about the possible usefulness of phrenology in predicting, preventing, and punishing crime. Combe republished the correspondence in an article (cited here) in *The Phrenological Journal*.

10. François Courvoisier murdered his master, Lord William Russell, in 1840.

11. Christine Ferguson refers to Peters's method as "affective detection," and contrasts it with the phrenologist's diagnosis, which she claims fails because it is visual rather than affective: "Peters first suspects North's guilt, not through the latter's deviant cranial stigmata, but through the hardness of expression and the coldness of speech . . . similarly, he reads Marwood's innocence, not in the regularity of his facial features, but in his spontaneous nervous reaction to the news of his uncle's murder" (10). Ferguson is the only critic to focus on the importance of Peters's unique method of detection, which can justifiably be described as "affective"—however, since Peters himself indicates his observations are insufficient to justify North's capture ("you can't give looks in as evidence" [246]), the phrenologist and Peters occupy similar, not opposed, positions in relation to detecting deviance and finding their techniques persuasively insufficient.

12. To be clear, O'Farrell's reading of the blush across a range of novels underscores its multivalent signifying functions, from being socially productive, to being erotically revealing, to being a site of resistance. In this respect the blush is far more ambiguous than Braddon's use of physiological indices of character.

13. Mangham argues that North's identification with the self-made man emerges in opposition to the "mid-Victorian ideas on pathological maternity" represented by the factory girl he jilts in the public house where Peters first encounters him (31). As he correctly observes, this is the crucial scene that allows Peters to connect North to the crime, but he attributes North's eventual arrest to the symbolic revenge of the pathologized mother. Here Mark Bennett makes a similar argument about North's criminality being based in a misrepresentation of his public identity, although he claims North's "true crime, in so far as it signifies within the world of Braddon's novel, has been the falsification of his own value"—that is, rising to the position of a rich European banker when he is really the product of an industrial town (44). Like Mangham, Bennett also identifies Peters's incidental encounter with North at the public house as the key moment in identifying the criminal, but focuses on the symbolic value of a coin thrown at North's face, an item that later aids in "the establishment of the villain's original 'coinage'" (44). Peters's attention to the coin, however, follows from his prior identification of North as the murderer from physiological signs.

14. C. Heywood makes the case for Hardy's and Eliot's indebtedness to *The Doctor's Wife* in "Miss Braddon's *The Doctor's Wife*: An Intermediary between *Madame Bovary* and *The Return of the Native*" and "A Source for *Middlemarch*: Miss Braddon's *The Doctor's Wife* and *Madame Bovary*," respectively.

15. For scholarship on the novel's relationship to *Madame Bovary*, see Edwards and Schaub. For readings about the novel's relationship to generic categories, see Beller and Sparks. For considerations of the novel's engagement with the effect of reading on the female reader, see Golden, Heilmann, Garrison, and Flint.

16. Notably, in the significant collection *Beyond Sensation: Mary Elizabeth Braddon in Context*, the dominant "context" is historical (whether social, legal, or political), yet the single essay on *The Doctor's Wife*, Tabitha Sparks's "Fiction Becomes Her: Representations of Female Character in Mary Braddon's *The Doctor's Wife*—is a consideration of the novel's ambivalent relationship to the genres of sensation, sentimental, and realist fiction.

17. Robert Lee Wolff identifies Sigismund Smith as an authorial "mouthpiece" (*Sensational* 126). Others who treat him as such include Pykett and Seys. Charles Raymond was first identified as Charles Bray by C. Heywood in "A Source for *Middlemarch*: Miss Braddon's *The Doctor's Wife* and *Madame Bovary*." In his biography of Braddon, Wolff concurs.

18. The entire second chapter of *The Philosophy of Necessity*, titled "Mind," is a lengthy elucidation of phrenological principles with frequent reference to Combe's works. Diana Postlethwaite offers an excellent summary and contextualization of this work in *Making It Whole* (122–32).

19. Similarly, the practical phrenologists Thorneycroft and Hallchurch observe, "more misery and unhappiness in this world results from the popular haphazard manner of choosing employment . . . youths are placed where they are not happy—they do not strive consequently, but rush to inevitable ruin" (3).

20. The author of *Phrenological Conversations* identifies himself as "An Uneducated Man," but as the book is published by James Deville, and as Deville was of humble origins and less formally educated than many of his fellow phrenologists, it seems reasonable to assume the author is Deville. See Roger Cooter's entry on Deville in *Phrenology in the British Isles*.

21. For instance, Catherine J. Golden argues that Isabel resembles Emma Bovary in the "way she reads novels with an uncritical acceptance" (33); Madeleine Seys says she is "neither a critical nor discerning reader" (178), and Ann Heilmann describes Isabel as having "infectious and narcissistic reading habits" (34).

22. Flint poses this question as one in a series that illustrate how women sensation writers often "leave one with moral, interpretative problems" that are not neatly answered by the narrative (292).

Chapter 5

1. For a detailed account of Eliot's relationship with the Brays, see chapter 4 of Kathryn Hughes's *George Eliot: The Last Victorian*.

2. In a letter to her friend Maria Lewis (*Letters* 1: 126), Eliot mentions having had her head read a second time. Eliot was thus phrenologically examined at least three times: twice before her letter to Lewis on 18 February 1842, and once by Deville in 1844, when the plaster cast of her skull was made. The first two readings were most likely conducted by Bray, whom she met in 1841. Eliot refers to Donovan's teachings and "wizard hand" in a letter to Mrs. Bray on 25 May 1845 (*Letters* 1: 193).

3. In the article, Lewes notes that "the overwhelming verdict of scientific authority is unequivocally against Phrenology . . ." (665). He attributes phrenology's widespread popularity in America and England to "active propagandism" (665) that panders to a general public that "buys treatises, attends lectures, collects skulls, and manipulates heads" (666).

4. Critics who have addressed Eliot's thoughts on phrenology tend to claim that her opinion of the science was largely positive (see Wright and Feltes). Critics who note Eliot's negative opinion of the science attribute her rejection to an increasing intimacy with George Henry Lewes, rather than her concerns with its social implications (Postlewaithe, *Making It Whole* 70; Gray 422).

5. In his diary entry for 20 October 1852, Combe notes, "Miss Marian Evans . . . has been our guest for a fortnight, & has left us this day" ("Continuous Diary: Britain and Abroad 1851–53").

6. Combe frequently mentions his many visits to Buckingham Palace to read and monitor the young prince's development in his journals. See "Continuous Diary: Britain and Abroad 1851–53," 20 and 29 May 1852; "Continuous Diary: Britain and Abroad 1853–54," 3 July 1853 and 27 June 1854.

7. See "Continuous Diary: Britain and Abroad 1851–53," 12 October 1852.

8. In a letter to Sara Hennell on 27 April 1846, Cara Bray mentions an interaction between Eliot and Simpson at their Rosehill home (qtd. in Eliot, *Letters* 1: 214 n.3).

9. For "The Lifted Veil's" use of sensory physiology, see Kennedy; for the story's connection to microsopy, see Xiao; for its correlation with mesmeric theory, see Bernstein ("Transatlantic Magnetism") and Ifill; for its engagement with telepathy, see Woods and Albrecht.

10. The only scholarship that gives more than glancing attention to the story's connection to phrenology is Beryl Gray's "Pseudoscience and George Eliot's 'The Lifted Veil,'" albeit mainly to pivot to the related subjects of mesmerism and clairvoyance.

11. Incidentally interesting given Latimer's claim to possess powers of telepathy and clairvoyance, an excess in the organ of wonder can also lead to an

unexamined belief in "fabulous narratives, in ghosts, inspirations, enchantments, and astrology" (*Elements* 104).

12. Eliot added the epigraph to the 1878 Cabinet Edition, which appeared a year before *Daniel Deronda*.

13. The Independent Section was discontinued after the first two numbers in 1852, so by the time it appeared, Combe's article did not appear in a separate section.

14. For reference to the lesson, see Eliot's letter to Mrs. Charles Bray (*Letters* 1: 193 n.3).

15. Sally Shuttleworth has also observed that "The Lifted Veil" and *Daniel Deronda* mirror one another, the latter work counteracting the former's "claustrophobic, pessimistic vision" in which "powers of prevision accorded to Mordecai become a powerfully enabling force" ("Introduction" xxviii).

16. See chapter 3 of *Realism, Photography, and Nineteenth-Century Fiction*. Galton's method consisted of exposing several portraits of individuals onto a single sensitized photographic plate in order to create a generalized image (Galton, "Composite Portraits").

17. Galton first used the term "eugenics" in his 1883 work, *Inquiries Into Human Faculty and Its Development* (25 n.1).

18. Eliot underlines the dramatic significance of Mrs. Meyrick's comment through Mirah's tearful reply, "'Oh please not to say that. . . . It is the first unkind thing you ever said'" (317).

19. Gilman describes this process of internalization in the Preface to *The Jew's Body*, speculating, "my sense is that the greater the identification of the Jew with the goals and values of the broader society, the more impacted the Jew is by the power of such images" (3).

20. The *Oxford English Dictionary* defines "emblem" as "A picture of an object (or the object itself) serving as a symbolical representation of an abstract quality, an action, state of things, class of persons."

21. Both George Levine and Jennifer Uglow identify Deronda's receptivity as an attribute linked with self-suppression, Levine noting its affinity to "the negative condition of the scientific mind" (24), and Uglow to "the negative capability of the artist's imagination" (234). This reading is less concerned with the negative quality itself than with the purpose for which Mordecai strategically claims it.

Afterword

1. In *The Blank Slate*, Pinker argues that, counter to most philosophic understandings of human nature, our behavior is largely innate and the result of evolutionary adaptations. Among humans' hardwired capacities he notes men's

tendency to excel at mental rotation, mathematics, and risk taking, whereas women are better at reading faces and showing concern for children.

2. Cordelia Fine coined the term "neurosexism" to characterize studies that appeal to neuroscience to explain behavioral differences between the sexes as the effect of innate and immutable psychological characteristics.

Works Cited

Abrams, M. H. *Natural Supernaturalism*. W. W. Norton and Co., 1971.
Albrecht, Thomas. "Sympathy and Telepathy: The Problem of Ethics in George Eliot's *The Lifted Veil*." *Victorian Literature and Culture*, vol. 45, no. 1, 2017, pp. 55–76.
Allan, J. McGrigor. "On the Real Differences in the Minds of Men and Women." *Journal of the Anthropological Society of London*, vol. 7, 1869, pp. 195–219.
Anti-Glorioso. "The Fire-away Style of Philosophy Briefly Examined and Illustrated." *The Zoist*, vol. 9, no, 33, 1851, pp. 65–69.
Armstrong, Mary. "Reading a Head: *Jane Eyre*, Phrenology, and the Homoerotics of Legibility." *Victorian Literature and Culture*, vol. 33, no. 1, 2005, pp. 107–32.
Armstrong, Nancy. *Desire and Domestic Fiction: A Political History of the Novel*. Oxford UP, 1987.
Atkinson, Henry George. "Mr. Atkinson on Mesmero-Phrenology." *Phrenological Journal and Magazine of Moral Science*, vol. 16, no. 74, 1843, pp. 326–28.
———. [On a family of idiots at Downham in Norfolk—to the London Phrenological Society.] *Zoist*, vol. 2, no. 6, 1844, pp. 163–85.
Atkinson, Henry George, and Harriet Martineau. *Letters on the Laws of Man's Nature and Development*, John Chapman, 1851.
Augustine. *Confessions*. Translated by Henry Chadwick, Oxford UP, 1998.
Austin, J. L. *How to Do Things with Words*. 2nd ed., Harvard UP, 1975.
Bain, Alexander. *On the Study of Character, Including an Estimate of Phrenology*. London, Parker, Son, and Bourn, 1861.
Barker, Juliet. *The Brontës: Wild Genius on the Moors: The Story of a Literary Family*. Pegasus Books, 2010.
Baron-Cohen, Simon. *The Essential Difference: Male and Female Brains and the Truth About Autism*. Basic Books, 2003.
Barwell, Louisa. *Nursery Government; Or, Hints Addressed to Mothers and Nursery-maids, on the Management of Young Children*. Chapman and Hall, 1836.

Beauvoir, Simone de. *The Second Sex*. Translated by Constance Borde and Sheila Malovary-Chevallier, Knopf, 2010.

Becker, Lydia. "Is There Any Specific Distinction Between Male and Female Intellect?" *The Englishwoman's Review*, vol. 8, 1868, pp. 483–91.

———. "On the Study of Science by Women." *Contemporary Review*, vol. 10, 1869, pp. 386–404.

Beer, Gillian. *Darwin's Plots: Evolutionary Narrative in Darwin, George Eliot and Nineteenth-Century Fiction*. 2nd ed., Cambridge UP, 2000.

Beller, Anne-Marie. "Sensational Bildung? Infantilization and Female Maturation in Braddon's 1860s Novels." *New Perspectives on Mary Elizabeth Braddon*, edited by Jessica Cox, Rodopi, 2012, pp. 113–31.

Bennett, Mark. "Generic Gothic and Unsettling Genre: Mary Elizabeth Braddon and the Penny Blood." *Gothic Studies*, vol. 13, no. 1, 2011, pp. 38–54.

Bernstein, Susan David. " 'Supposed Differences': Lydia Becker and Victorian Women's Participation in the BAAS." *Repositioning Victorian Sciences: Shifting Centres in Nineteenth-Century Scientific Thinking*, edited by David Clifford, Elisabeth Wadge and Alex Warwick, Anthem, 2006, pp. 85–94.

———. "Transatlantic Magnetism: Eliot's 'The Lifted Veil' and Alcott's Sensation Stores." *Transatlantic Sensations*, edited by Jennifer Phegley, John Cyril Barton, and Kristin N. Huston, Ashgate, 2012, pp. 183–206.

Berry, Laura. "Acts of Custody and Incarceration in *Wuthering Heights* and *The Tenant of Wildfell Hall*." *Novel*, vol. 30, no. 1, fall 1996, pp. 32–55.

Bjaalie, Jan G., Shigeo Okabe, and Linda J. Richards. "International Brain Initiative: An Innovative Framework for Coordinated Global Brain Research Efforts." *Neuron*, vol. 105, no. 2, 2020, pp. 212–16. http://doi.org/10.1016/j.neuron.2020.02.022

Bodenheimer, Rosmarie. *The Real Life of Mary Ann Evans: George Eliot, Her Letters and Fiction*. Cornell UP, 1994.

Bordo, Susan. *Unbearable Weight: Feminism, Western Culture, and the Body*. U of California P, 1993.

Braddon, Mary Elizabeth. *Birds of Prey*. 4th ed., Ward, Lock, and Tyler, 1867, 4 vols.

———. *The Doctor's Wife*. Edited by Lyn Pykett, Oxford UP, 2008.

———. *Lady Audley's Secret*. Edited by Lynn Pykett, Oxford UP, 1998.

———. "My First Novel." *The Trail of the Serpent*, Mary Elizabeth Braddon. Edited by Chris Willis, Random House, 2003, pp. 415–27.

———. *The Trail of the Serpent*. Edited by Chris Willis, Random House, Inc., 2003.

Bray, Charles. *Education of the Feelings*, Taylor and Walton, 1838.

———. *Phases of Opinion and Experience During a Long Life: An Autobiography*. Longmans, Green, and Co., 1884.

―――. *The Philosophy of Necessity; or, The Law of Consequences; as Applicable to Mental, Moral, and Social Science*. Longman, Orme, Brown, Green, and Longmans, 1841.
Bridges, F. *A Descriptive Catalogue of the Casts, Busts, Masks, Skulls and Drawings, in F. Bridges' Phrenological Gallery*. Joshua Hobson, 1838.
[Broca, Paul]. "Broca on Anthropology." *Anthropological Review*, vol. 6, 1868, pp. 35–52.
Brontë, Anne. *Agnes Grey*. Edited by Robert Inglesfield and Hilda Marsden, Oxford UP, 2010.
―――. *The Tenant of Wildfell Hall*. Edited by Lee A. Talley, Broadview Press, 2009.
Brontë, Charlotte. *Jane Eyre*. Edited by Richard J. Dunn, 3rd ed., W. W. Norton, 2001.
―――. *The Professor*. Edited by Margaret Smith and Herbert Rosengarten, Oxford UP, 2008.
Brontë, Patrick. Letter to Harriet Martineau. 5 Nov. 1857, Harriet Martineau Papers, Cadbury Research Library, U of Birmingham, Manuscripts and Archives, HM 90.
Brooks, Peter. *Troubling Confessions*. U of Chicago P, 2000.
Browne, James. *Phrenology and Its Application to Education, Insanity, and Prison Discipline*. Bickers and Sons, 1869.
Bunyan, John. *The Pilgrim's Progress*. Edited by James Reeves, Penguin, 1987.
Bushnan, J. Stevenson. *Miss Martineau and Her Master*. John Churchill, 1851.
Butler, Judith. *Gender Trouble: Feminism and the Subversion of Identity*. Routledge, 1990.
Caldwell, Charles. "Thoughts on the Pathology, Prevention and Treatment of Intemperance, as a Form of Mental Derangement." *The Transylvania Journal of Medicine and the Associate Sciences*, vol. 5, no. 3, 1832, pp. 309–50.
Capuano, Peter. *Changing Hands: Industry, Evolution, and the Reconfiguration of the Victorian Body*. U of Michigan P, 2015.
Carmichael, Andrew. *A Memoir of the Life and Philosophy of Spurzheim*. Marsh, Caper, Lyon, Lilly, Wait and Co., 1833.
Carnell, Jennifer. *The Literary Lives of Mary Elizabeth Braddon: A Study of Her Life and Work*. Sensation Press, 2000.
Carpenter, Mary Wilson. *George Eliot and the Landscape of Time*. U of North Carolina P., 1986.
Carroll, Joseph. *Literary Darwinism: Evolution, Human Nature, and Literature*. Taylor & Francis, 2004.
"Case of Courvoisier, lately executed for the Murder of his Master, Lord William Russell." *The Phrenological Journal*, vol. 13, no. 65, 1840, 323–41.
"Case of Ehlert, the Prussian mate, lately executed for the murder of Berckholtz. . . ." *The Phrenological Journal*, vol. 13, no. 62, 1840, pp. 59–79.

"Case of William Saville, executed at Nottingham for Murder." *Phrenological Journal*, vol. 17, no. 79, 1844, pp. 385–92.

Cawelti, John G. *Adventure, Mystery, and Romance: Formula Stories as Art and Popular Culture*. U of Chicago P, 1976.

Chyette, Bryan. *Constructions of "The Jew" in English Literature and Society: Racial Representations, 1875–1945*. Cambridge UP, 1993.

Claggett, Shalyn. "Putting Character First: The Narrative Construction of Innate Identity in Phrenological Texts." *Victorians Institute Journal*, vol. 38, 2010, pp. 103–26.

Clarke, Henry. *Christian Phrenology; Or, The Teachings of the New Testament Respecting the Animal, Moral, and Intellectual Nature of Man*. J. Anderson, 1835.

Clayton, Jay. *Charles Dickens in Cyberspace: The Afterlife of the Nineteenth Century in Postmodern Culture*. Oxford UP, 2003.

———. *Romantic Vision and the Novel*. Cambridge UP, 1987.

Cobbe, Frances Power. "Feminine Brains." *The Englishwoman's Review*, vol. 19, March 1888, pp. 111–12.

———. *Life of Frances Power Cobbe, By Herself*. Richard Bentley & Son, 1894.

Combe, George. *The Constitution of Man Considered in Relation to External Objects*. 7th ed., MacLachlan, Stewart, and Co., 1836.

———. "Continuous Diary: Britain and Abroad: 1851–53." 29 May 1852, National Library of Scotland, Archives and Manuscripts, MS 7428 folio 47.

———. "Continuous Diary: Britain and Abroad: 1851–53." 12 Oct. 1852, National Library of Scotland, MS 7428, folio 111.

———. "Continuous Diary: Britain and Abroad: 1851–53." 14 Oct. 1852, National Library of Scotland, Archives and Manuscripts, MS 7428, folio 112.

———. "Continuous Diary: Britain and Abroad: 1851–53." 16 Oct. 1852, National Library of Scotland, Archives and Manuscripts, MS 7428, folio 113.

———. "Continuous Diary: Britain and Abroad: 1851–53." 20 Oct. 1852, National Library of Scotland, Archives and Manuscripts, MS 7428, folio 113.

———. "Continuous Diary: Britain and Abroad: 1853–54." 3 July 1853, National Library of Scotland, Archives and Manuscripts, MS 7429, folio 22.

———. "Continuous Diary: Britain and Abroad: 1853–54." 27 June 1854, National Library of Scotland, Archives and Manuscripts, MS 7429, folio 107.

———. "Criminal Legislation and Prison Discipline." *The Westminster Review*, vol. 16, 1854, pp. 409–45.

———. *Elements of Phrenology*. 7th ed., Maclachlan, Stewart, and Co., 1850.

———. *Lectures on Popular Education; Delivered to the Edinburgh Philosophical Association*. 3rd ed., MacLachlan, Stewart, and Co., 1848.

———. "Letter from George Combe to Francis Jeffrey, Esq." *Phrenological Journal*, vol. 4, no. 13, 1827, pp. 1–82.

———. Letter to Henry George Atkinson. 19 Oct. 1851, The Combe Papers, National Library of Scotland, Manuscripts and Archives, MS 7392, folio 407.

———. *Moral Philosophy; Or, The Duties of Man Considered in his Individual, Domestic, and Social Capacities.* 3rd ed., MacLachlan, Steward, and Co., 1846.
———. *System of Phrenology.* Harper & Brothers, 1860.
"The 'Coming Out' of Woman." *Phrenological Journal and Science of Health,* vol. 103, no. 3, 1897, p. 109.
Cooter, Roger. *The Cultural Meaning of Popular Science: Phrenology and the Organization of Consent in Nineteenth-century Britain.* Cambridge UP, 1984.
———. *Phrenology in the British Isles: An Annotated, Historical Biobibliography and Index.* Scarecrow Press, 1989.
C.T.D. Advertisement for *Remembrances of a Monthly Nurse* by Harriet Downing. Harriet Downing, *Rememberances of a Monthly Nurse,* Simms and M'Intyre, 1852.
Dames, Nicholas. *Amnesiac Selves: Nostalgia, Forgetting, and British Fiction, 1810–1870.* Oxford UP, 2001.
———. "The Clinical Novel: Phrenology and *Villette.*" *Novel,* vol. 29, no. 3, 1996, pp. 367–90.
———. *The Physiology of the Novel: Reading, Neural Science, and the Form of Victorian Fiction.* Oxford UP, 2007.
Darwin, Charles. *The Descent of Man, and Selection in Relation to Sex.* John Murray, 1874.
De Giustino. *Conquest of Mind: Phrenology and Victorian Social Thought.* Croom Helm, 1975.
Deville, James. "Account by Mr. Deville of his Phrenological Collection in London." *The Phrenological Journal,* vol. 14, no. 66, 1841, pp. 19–23.
———. *Phrenological Conversations, Introductory to an Inquiry as to the Establishment of Phrenology as a Science with Some Illustrations of Its Application in Education, Insanity, Convict Discipline, and Social Life.* J. Deville, 1833.
Dewhirst, Ian. "The Rev. Patrick Brontë and the Keighley Mechanics' Institute." *Brontë Society Transactions,* vol. 14, no. 5, 1965, pp. 35–37.
Dexter, S. "Education." *Phrenological Journal and Science of Health,* vol. 104, no. 4, 1897, pp. 162–64.
Diprose, Rosalyn. *The Bodies of Women: Ethics, Embodiment and Sexual Difference.* Routledge, 1994.
Dodds, Susanna W. "A Healing Art in the Twentieth Century." *Phrenological Journal and Science of Health,* vol. 104, 1897, pp. 175–76.
[Downing, Harriet.] "Remembrances of a Monthly Nurse. Lord Walter Maxwell." *Fraser's Magazine,* vol. 16, 1837, pp. 497–512.
Ecker, Alexander. "On a Characteristic Peculiarity in the Form of the Female Skull, and Its Significance for Comparative Anthropology." *Anthropological Review,* vol. 6, no. 23, 1868, pp. 350–56.
Edwards, P. D. "French Realism Englished: The Case of M. E. Braddon's *The Doctor's Wife.*" *Victorian Turns, NeoVictorian Returns: Essays on Fiction and*

Culture, edited by Penny Gay, Judith Johnston, and Catherine Waters, Cambridge Scholars, 2008, pp. 113–23.

Eliot, George. *Daniel Deronda*. Edited by Graham Handley, Oxford UP, 1998.

———. *The George Eliot Letters*. Edited by Gordon Haight, Yale UP, 1954–78, 9 vols.

———. "The Lifted Veil." *Blackwood's Magazine*, vol. 86, July 1859, pp. 24–48.

———. *The Lifted Veil and Brother Jacob*. Edited by Sally Shuttleworth, Penguin, 2001, pp. 1–42.

———. "A Minor Prophet." *George Eliot: Collected Poems*, edited by Lucien Jenkins, Skoob Books, 1989, pp. 31–40.

———. "A Word for the Germans." *The Essays of George Eliot*, edited by Thomas Pinney. New Colombia University Press, 1963, pp. 386–90.

———. "Woman in France: Madame de Sablé." *The Essays of George Eliot*, edited by Thomas Pinney, New Colombia University Press, 1963, pp. 52–81.

Eliot, Lise. "The Trouble with Sex Differences." *Neuron*, vol. 72, no. 6, 2011, pp. 895–98. http://doi.org/10.1016/j.neuron.2011.12.001

Elliott, Nathan R. "Phrenology and the Visual Stereotype in Charlotte Brontë's *Villette*." *Nineteenth Century Studies*, vol. 22, 2008, pp. 41–55.

Elliotson, [John]. "An Account of the Head of Rush, the Norfolk Murderer." *The Zoist*, vol. 7, no. 26,1849, pp. 107–21.

Ellis, Ida Mitchel. *How to Improve Body, Brain, and Mind*. 3rd ed., Ellis Family, 1904.

Ellis, [Sarah Stickney]. *The Daughters of England, Their Position in Society, Character and Responsibilities*. Fischer, Son, and Co., 1842.

"emblem, n." *The Oxford English Dictionary*. 2nd ed., Oxford UP, 1989.

Farnes, Patricia. "Women in Medical Science." *Women of Science: Righting the Record*, edited by G. Kass-Simon and Patricia Farnes, Indiana UP, 1993, pp. 268–99.

Feltes, N. N. "Phrenology: from Lewes to George Eliot." *Studies in the Literary Imagination*, vol. 1, no. 1, 1968, pp. 13–22.

"Female Astronomers." *Phrenological Magazine*, vol. 1, no. 10, 1885, pp. 419–24.

Ferguson, Christine. "Sensational Dependence: Prosthesis and Affect in Dickens and Braddon." *LIT: Literature Interpretation Theory*, vol. 19, no. 1, 2008, pp. 1–25.

"A Few Words Upon the Life and Writings of the Late Mrs. Harriet Downing." *Metropolitan Magazine*, vol. 43, 1845, pp. 94–96.

Fine, Cordelia. *Delusions of Gender: How Our Minds, Society, and Neurosexism Create Difference*, W. W. Norton, 2010.

Flint, Kate. *The Woman Reader: 1837–1914*. Clarendon Press, 1993.

Forbes, John. "Phrenology." *British and Foreign Medical Review*, vol. 9, no. 17, 1840, pp. 190–215.

Foucault, Michel. *Discipline and Punish: The Birth of the Prison*. Translated by Alan Sheridan, Vintage Books, 1995.

———. *The History of Sexuality*. Translated by Robert Hurley, vol. 1, Random House, 1978.

———. *Power/Knowledge: Selected Interviews and Other Writings: 1872–1977*. Edited by Colin Gordon, translated by Colin Gordon, et al, Vintage Books, 1980.

Fowler, Jessie A. *A Manual of Mental Science for Teachers and Students of Childhood: Its Character and Culture*. L. N. Fowler, 1897.

———. *Men and Women Compared; Or, Their Mental and Physical Differences Considered*. L. N. Fowler, n.d.

Fowler, L[orenzo] N[iles]. "Phrenology in England." *The Phrenological Magazine*, vol. 1, 1880, pp. 5–7.

Fowler, Lydia. *Woman, Her Destiny and Maternal Relations; or, Hints to the Single and Married*. William Tweedie, 1864.

Frawley, Maria H. *A Wider Range: Travel Writing by Women in Victorian England*. Fairleigh Dickinson UP, 1994.

Friedan, Betty. *The Feminine Mystique*. W. W. Norton, 2001.

Froude, James Anthony. "Materialism.—Miss Martineau and Mr. Atkinson." *Fraser's*, vol. 43, 1851, pp. 418–34.

Fuss, Diane. *Essentially Speaking: Feminism, Nature, and Difference*. Routledge, 1989.

Galton, Francis. *Hereditary Genius: An Inquiry Into Its Laws and Consequences*. Macmillan and Co., 1914.

———. *Inquiries into Human Faculty and Development*. Macmillian, 1883.

———. "Photographic Composites." *Photographic News*, vol. 29, 1885, pp. 243–45.

Garrison, Laurie. "'She read on more eagerly, almost breathlessly': Mary Elizabeth Braddon's Challenge to Medical Depictions of Female Masturbation in *The Doctor's Wife*." *The Female Body in Medicine and Literature*, edited by Andrew Mangham and Greta Depledge, Liverpool UP, 2011.

Gaskell, Elizabeth. *The Life of Charlotte Brontë*. Penguin, 1985.

Gates, Barbara T. "Ordering Nature: Revisioning Victorian Science Culture." *Victorian Science in Context*, edited by Bernard Lightman, U of Chicago P, 1997, pp. 179–86.

"George Combe." *The English Women's Journal*, vol. 2., no. 7, 1858, pp. 53–56.

G.F.S. "Individuality of Woman." *Phrenological Magazine*, vol. 2, no. 7, 1858, pp. 53–56.

Gibbon, Charles. *The Life of George Combe*. Vol. 1, Macmillian, 1878.

Gieryn, Thomas F. *Cultural Boundaries of Science: Credibility on the Line*. Chicago UP, 1999.

Gilman, Sander L. "Black Bodies, White Bodies: Toward an Iconography of Female Sexuality in Late Nineteenth-Century Art, Medicine, and Literature." *Critical Inquiry*, vol. 12, no. 1, 1985, pp. 204–42.

———. *The Jew's Body*. Routledge, 1991.

Ginzburg, Carlo. "Family Resemblances and Family Trees: Two Cognitive Metaphors." *Critical Inquiry*, vol. 30, no. 3, 2004, pp. 537–56.

Golden, Catherine. "Censoring Her Senationalism: Mary Elizabeth Braddon and *The Doctor's Wife*." *Victorian Sensations: Essays on a Scandalous Genre*, edited by Kimberly Harrison and Richard Fantina, Ohio State UP, 2006, pp. 29–40.

Gould, Stephen Jay. *The Mismeasure of Man*. W. W. Norton, 1981.

Gray, B. M. "Pseudoscience and George Eliot's 'The Lifted Veil.'" *Nineteenth-Century Fiction*, vol. 36, no. 4, 1982, pp. 407–23.

Green, James Aaron. "Seeing in the City: Modern Visuality in M. E. Braddon's *The Trail of the Serpent*." *Victorian Network*, vol. 9, 2020, pp. 80–100.

Grosz, Elizabeth. *Volatile Bodies: Toward a Corporeal Feminism*. Indiana UP, 1994.

Guillot, Edouard. "The Phrenological Constitution of Woman." *Journal of Health*, vol. 6, 1857, pp. 193–94.

Gurain, Michael. *What Could He Be Thinking?: How a Man's Mind Really Works*. St. Martin's Griffin, 2003.

Hall, Spencer. Introduction. *The Phreno-Magnet, and Mirror of Nature*, vol. 1, no. 1, 1843, pp. 1–3.

Hallchurch T., and W. Thorneycroft, *The New Phrenological and Physiolgnomical Chart (Embodying Trades)*. Oldbury and West Bromwich, 1884.

Haraway, Donna J. *Simians, Cyborgs, and Women: The Reinvention of Nature*. Routledge, 1991.

Harcourt, A. V. Letter to Harriet Martineau. 4 Jan. 1856, Harriet Martineau Papers, Cadbury Research Library, U of Birmingham, Manuscripts and Archives, HM 423.

Harley, Alexis. *Autobiologies: Charles Darwin and the Natural History of the Self*. Bucknell UP, 2014.

Hartley, Lucy. "A Science for One or a Science for All? Physiognomy, Self-Help and the Practical Benefits of Science." *Repositioning Victorian Sciences: Shifting Centres in Nineteenth-Century Scientific Thinking*, edited by David Clifford et al., Anthem P, 2006, pp. 71–82.

Heilmann, Ann. "Emma Bovary's Sisters: Infectious Desire and Female Reading Appetites in Mary Braddon and George Moore." *Victorian Review*, vol. 29, no. 1, 2003, pp. 31–48.

Heywood, C. "Miss Braddon's *The Doctor's Wife*: An Intermediary between *Madame Bovary* and *Return of the Native*." *Revue de Littérature Compare*, vol. 38, 1964, pp. 255–61.

———. "A Source for *Middlemarch*: Miss Braddon's *The Doctor's Wife* and *Madame Bovary*." *Revue de Littérature Compare*, vol. 44, 1970, 184–94.

The History of Woman Suffrage, 1848–1861. Edited by Elizabeth Cady Stanton, Susan B. Anthony, and Matilda Joslyn Gage, vol. 1, Fowler and Wells, 1881.

Hoecker-Drysdale, Susan. "The Enigma of Harriet Martineau's Letters on Science." *Women's Writing*, vol. 2, no. 2, 1995, pp. 155–65.

———. *Harriet Martineau: First Woman Sociologist*. Berg, 1992.
"Homicidal Monomania and Murder by Premedication." *Lancet*, vol. 10, no. 1032, 1842, pp. 376–77.
Hughes, Kathryn. *George Eliot: The Last Victorian*. Farrar, Straus, and Giroux, 1998.
Hunt, Harriet K. *Glances and Glimpses; Or Fifty Years Social, Including Twenty Year Professional, Life*. John P. Jewett and Co., 1856.
Hunter, Robert. *The Philosophy of Phrenology Simplified*. W. R. M'Phun, 1838.
Hunter, Shelagh. *Harriet Martineau: The Poetics of Moralism*. Scholar Press, 1995.
Huxley, Thomas H. *Science and Education: Essays by Thomas H. Huxley*, vol. 3, Appleton and Co., 1897.
Hyde, Janet, "The Gender Similarities Hypothesis." *American Psychologist*, vol. 60, no. 6, pp. 581–92.
Hyman, Gwen. "'An infernal fire in my veins': Gentlemanly Drinking in *The Tenant of Wildfell Hall*. *Victorian Literature and Culture*, vol. 36, no. 2, 2008, pp. 451–69.
Hytche, E. J. "On the Selection of Keepers in Lunatic Asylums." *The Lancet*, vol. 1, 1841, pp. 190–92.
Ifill, Helena. "Mesmeric Clairvoyance in Mid-Victorian Literature: Eliot, Bulwer-Lytton, and MacDonald." *Supernatural Studies*, vol. 2, no. 2, 2015, pp. 118–32.
Jack, Ian. "Physiognomy, Phrenology and Characterisation in the Novels of Charlotte Brontë." *Brontë Society Transactions*, vol. 16, no. 5, 1970, pp. 377–91.
Jacobs, N. M. "Gender and Layered Narrative in *Wuthering Heights* and *The Tenant of Wildfell Hall*." *The Brontës*, edited by Patricia Ingham London, 2003, pp. 216–33.
James, Henry. "*Daniel Deronda*: A Conversation." *The Atlantic Monthly*, vol. 38, 1876, pp. 684–94.
"Jessie A. Fowler, Phrenologist, Dies." *New York Times*, 16 Oct. 1932, p. 38.
Joel, Daphna, et al. "Sex Beyond the Genitalia: The Human Brain Mosaic." *Proceedings of the National Academy of Sciences*, vol. 112, no. 50, 2015, pp. 15468–73. https://doi.org/10.1073/pnas.150965411
Jones, Ann Rosalind. "Writing the Body: Toward an Understanding of *l'écriture féminine*." *Feminist Studies*, vol. 7, no. 2, 1981, pp. 247–63.
Jordan-Young, Rebecca M. *Brainstorm: The Flaws in the Science of Sex Differences*. Harvard UP, 2010.
Joshi, Priti. "Masculinity and Gossip in Anne Brontë's *Tenant*." *SEL*, vol. 49, no. 4, 2009, pp. 907–24.
Kennedy, Meegan. "'A True Prophet'? Speculation in Victorian Sensory Physiology and George Eliot's 'The Lifted Veil.'" *Nineteenth-Century Literature*, vol. 71, no. 2, 2016, pp. 360–403.
Kersey, Alyssa J., Kelsey D. Csumitta, and Jessica F. Cantlon. "Gender Similarities in the Brain During Mathematics Development." *Science of Learning*, vol. 4, no. 19, 2019, pp. 1–19.

Ketabgian, Tamara. "Martineau, Mesmerism, and the 'Night Side of Nature." *Women's Writing*, vol. 9, no. 3, 2002, pp. 351–68.
[Kingsley, Charles]. "Recent Novels." *Fraser's Magazine for Town and Country*, vol. 39, 1849, pp. 417–32.
Knoepflmacher, U. C. *Religious Humanism and the Victorian Novel*. Princeton UP, 1965.
Kurshan, Ilana. "Mind Reading: Literature in the Discourse of Early Victorian Phrenology and Mesmerism." *Victorian Literary Mesmerism*, edited by Martin Willis and Catherine Wynne, Rodopi, 2006, pp. 17–38.
Laquer, Thomas. *Making Sex: Body and Gender from the Greeks to Freud*. Cambridge: Harvard UP, 1990.
Lee, Alice. "Data for the Problem of Evolution in Man." *Philosophical Transactions of the Royal Society*, vol. 196, 1901, pp. 225–64.
Lersner, Elise Von. *Children's Gifts and Mothers' Duties: or a Book for Mothers*. G. E. Waters, [1865].
Levine, George. *Darwin and the Novelists: Patterns of Science in Victorian Fiction*. U of Chicago P, 1991.
———. "George Eliot's Hypothesis of Reality." *Nineteenth-Century Fiction*, vol. 35, no. 1, 1980, pp. 1–28.
[Lewes, George Henry and John F. W. Herschel]. "Notes on Science." *Cornhill Magazine*, vol. 7, 1863, pp. 276–80.
Lewes, George Henry. *The History of Philosophy from Thales to Comte*. 3rd ed., Longmans, Green, and Co., 1867. 2 vols.
———. "Phrenology in France." *Blackwood's Magazine*, vol. 82, 1857, pp. 665–74.
Logan, Deborah. *The Hour and the Woman: Harriet Martineau's 'Somewhat Remarkable' Life*. Northern Illinois UP, 2002.
Lowe, W. R. "Case of John Williams, executed at Shrewsbury, on Saturday, April 2. 1842, for the willful Murder of Emma Evans, at Chirk, near Oswestry, Salop." *Phrenological Journal*, vol. 16, no. 74, 1843, pp. 62–70.
"Lydia Folger Fowler." *The Englishwoman's Review of Social and Industrial Questions*, vol. 10, 1879, pp. 82–83.
Lyons, J. C. *The Science of Phrenology, as Applicable to Education, Friendship, Love, Courtship, and Matrimony, Etc.* Aylott and Jones, 1846.
Mackenzie, G. S. "[Testimonial] From Sir G. S. Mackenzie, Bart. F.R.S.L., formerly President of the Physical Class of the Royal Society of Edinburgh. . . ." *Testimonials on Behalf of George Combe, as a Candidate for the Chair of Logic in the University of Edinburgh*. John Anderson, Longman and Co, 1836, pp. 7–8.
Macnish, Robert. *The Anatomy of Drunkenness*. W. R. M'Phun, 1859.
———. *The Philosophy of Sleep*. N. H. Cotes, 1838.
———. "[Testimonial] From Robert Macnish, Esq. Member of the Faculty of Physicians and Surgeons of Glasgow. . . ." *Testimonials on Behalf of George*

Combe, as a Candidate for the Chair of Logic in the University of Edinburgh, John Anderson, Longman and Co., 1836, pp. 15–17.

Mangham, Andrew. "'Drink it up dear; it will do you good': Crime, Toxicology, and *The Trail of the Serpent*." *New Perspectives on Mary Elizabeth Braddon*, edited by Jessica Cox, Rodopi, 2012.

———. "'Murdered at the Breast': Maternal Violence and the Self-Made Man in Popular Victorian Culture." *Critical Survey*, vol. 16, no. 1, 2004, pp. 20–34.

"Mares'-Nests: Found by the Materialists, the Owenites, and the Craniologists." *Fraser's Magazine*, vol. 9, 1834, pp. 424–34.

Martineau, Harriet. *Biographical Sketches by Harriet Martineau*. Leypoldt and Holt, 1869.

———. *The Collected Letters of Harriet Martineau*. Pickering and Chatto, 2007. 5 vols.

———. *Daily News* [London]. 2 April 1856, p. 4.

———. *Eastern Life, Present and Past*. Lea and Blanchard, 1848.

———. "Female Industry." *Edinburgh Review*, vol. 109, 1859, pp. 293–336.

———. *Harriet Martineau's Autobiography*. Vol. 2, 3rd ed., Smith, Elder, and Co., 1877, 2 vols.

———. Codicil to the Last Will and Testament of Harriet Martineau. 1872, Harriet Martineau Papers, Cadbury Research Library, U of Birmingham, Manuscripts and Archives, HM 679.

———. Last Will and Testament of Harriet Martineau. 1864, Harriet Martineau Papers, Cadbury Research Library, U of Birmingham, Manuscripts and Archives, HM 679.

———. *Letters on Mesmerism*. Edward Moxon, 1845.

———. "On Female Education." *Monthly Repository*, vol. 18, 1823, pp. 77–81.

———. *Selected Letters*. Edited by Valerie Sanders, Clarendon Press, 1990.

———. *Society in America*. Saunders and Otley, 1837, 3 vols.

———. "What Women are Educated For." *Once a Week*, vol. 5, 1861, pp. 175–79.

Martineau, James. "Mesmeric Atheism." *Prospective Review*, vol. 7, 1851, pp. 224–62.

"Materialism and the Phrenological Association." *Phrenological Journal*, vol. 16, no. 74, 1843, pp. 41–59.

May, Leila S. "Lucy Snowe, a Material Girl? Phrenology, Surveillance, and the Sociology of Interiority." *Criticism*, vol. 55, no. 1, 2013, pp. 43–68.

McDonagh, Josephine. Introduction. *The Tenant of Wildfell Hall*, Anne Brontë, Oxford UP, 2008, pp. i–xxxv.

McKay, Brenda. "Victorian Anthropology and Hebraic Apocalyptic Prophecy: 'The Lifted Veil.'" *George Eliot-George Henry Lewes Studies*, vol. 42–43, 2002, pp. 69–92.

McKee, Patricia. "Racial Strategies in *Jane Eyre*." *Victorian Literature and Culture*, vol. 37, no. 1, 2009, pp. 67–83.

Meyer, Susan. *Imperialism at Home: Race and Victorian Women's Fiction.* Cornell UP, 1996.
Miles, Mrs. L. *Phrenology, and the Moral Influence of Phrenology.* Carey, Lea and Blanchard, 1835.
Miller, Mrs. F. Fenwick. *Harriet Martineau.* W. H. Allen & Co., 1884.
"Miss Becker on the Mental Characteristics of the Sexes." *The Lancet,* vol. 2, 1868, pp. 320–21.
"Miss Caroline Hazard." *Phrenological Journal and Science of Health,* vol. 108, no. 12, 1899, p. 384.
Mittermair, C. J. A., and George Combe. "On the Application of Phrenology to Criminal Legislation and Prison Discipline." *Phrenological Journal,* vol. 16, no. 74, 1843, pp. 1–19.
Mohammadi, Dara. "ENIGMA: Crowdsourcing Meets Neuroscience." *The Lancet,* vol. 14, no. 5, 2015, pp. 462–63. https://doi.org/10.1016/S1474-4422(15)00005-8
Montgomery, James. "An Essay on the Phrenology of the Hindoos and Negroes." *Phrenological Journal,* vol. 6, no. 24, 1830, pp. 398–410.
Morrell, Jack, and Arnold Thackray. *Gentlemen of Science: Early Years of the British Association for the Advancement of Science,* Royal Historical Society, 1984.
Morse, Deborah Denenholz. "'I speak of those I do know': Witnessing as Radical Gesture in *The Tenant of Wildfell Hall.*" *New Approaches to the Literary Art of Anne Brontë,* edited by Julie Nash and Barbara A. Suess, Ashgate, 2001, pp. 103–26.
Moscucci, Ornella. *The Science of Woman: Gynaecology and Gender in England, 1800–1929.* Cambridge UP, 1990.
"Mr. George Combe and the Philosophy of Phrenology." *Fraser's Magazine,* vol. 22, no. 131, 1840, pp. 509–20.
"Mrs. Emily Crawford." *Phrenological Journal and Science of Health,* vol. 106, no. 2, 1898, pp. 45–47.
"Mrs. Hamilton." *The Crisis,* vol. 3, 1834, p. 32.
"Mrs. James Fairman." *Phrenological Journal and Science of Health,* vol. 106, no. 2, 1898, pp. 45–47.
"Mrs. Mary Lyon Dame Hall." *Phrenological Journal and Science of Health,* vol. 104, no. 1, 1897, pp. 13–18.
"Mrs. Oliphant." *Phrenological Journal and Science of Health,* vol. 103, no. 4, 1897, pp. 158–60.
Murphy, James. *Church, State, and Schools in Britain, 1800–1970.* Routledge and Keegan Paul, 1971.
Murphy, Patricia. *In Science's Shadow: Literary Constructions of Late Victorian Women.* U of Missouri P, 2006.
New International Version Bible. Kenneth Barker, general editor, Zondervan Publishing House, 1995.

Novak, Daniel A. *Realism, Photography, and Nineteenth-Century Fiction*. Cambridge UP, 2008.
O'Dell, Stackpool E. and Geelossapuss O'Dell. *Phrenology: Essays and Studies*. London Phrenological Institution, 1899.
O'Farrell, Mary Ann. *Telling Complexions: The Nineteenth-century English Novel and the Blush*. Duke UP, 1997.
[Oliphant, Margaret.] "Novels." *Blackwood's Magazine*, vol. 102, no. 2, 1867, pp. 257–80.
"On the Development and Character of the North American Indians." *Phrenological Journal*, vol. 2, no. 8, 1825, pp. 533–43.
Oppenheim, Annie I. *Phreno-Physiology: Scientific Character-Reading from the Face. Illustrated by Itala Buccani*. Simpkin, Marshal & Co., 1892.
Otto, [Carl]. "Remarkable Case of Sudden Morbid Excitement of Destructiveness, depending upon a bodily disease." *Phrenological Journal*, vol. 12, no. 60, 1839, pp. 255–56.
Owen, Alex. *The Darkened Room: Women, Power and Spiritualism in Late Victorian England*. U of Pennsylvania P, 1990.
Parker, Joan E. "Lydia Becker's 'school for science': A Challenge to Domesticity." *Woman's History Review*, vol. 10, no. 4, 2001, pp. 629–50.
Parker, Margaret E. "Healthy Work for Women." *The Englishwoman's Review*, vol. 12, no. 99, 1881, pp. 295–99.
"Paper Read at the First Meeting of the Hastings Branch." *Phrenological Magazine*, vol. 4, 1888, pp. 163–64.
Parssinen, Terry M. "Professional Deviants and the History of Medicine: Medical Mesmerists in Victorian Britain." *On the Margins of Science: The Social Construction of Rejected Knowledge*, edited by Roy Wallis, Sociological Review Monograph 27, 1979, pp. 103–20.
Patenall, Annie N. "Character." *Phrenological Magazine*, vol. 5, 1889, pp. 228–37.
———. "Paper Read at the First Meeting of the Hastings Branch." *Phrenological Magazine*, vol. 4, no. 40, 1881, pp. 163–64.
———. "To the Editor of the Observer." *Phrenological Magazine*, vol. 5, no. 57, 1889, p. 387.
Peterson, Linda H. *Traditions of Victorian Women's Autobiography: The Poetics and Politics of Life Writing*. UP of Virginia, 1999.
"Phrenology Applied in Education." *Phrenological Journal*, vol. 6, 1830, pp. 238–44.
Pichanick, Valerie Kossew. *Harriet Martineau: The Woman and Her Work*. U of Michigan P, 1980.
Pike, Luke Owen. "On the Claims of Women to Political Power." *Journal of the Anthropological Society of London*, vol. 7, 1869, pp. 47–61.
Pinker, Steven. *The Blank Slate: The Modern Denail of Human Nature*. Penguin, 2002.

———. "Sex Ed: The Science of Difference." *The New Republic*, 14 Feb. 2005, pp. 15–17.

Poovey, Mary. *Uneven Developments: The Ideological Work of Gender in Mid-Victorian England*. U of Chicago P, 1988.

"Pope versus Phrenology." *Phrenological Journal*, vol. 10, no. 52, 1837, pp. 600–2.

Postlethwaite, Diana. *Making It Whole: A Victorian Circle and the Shape of Their World*. Ohio State UP, 1984.

———. "Mothering and Mesmerism in the Life of Harriet Martineau." *Signs*, vol. 14, no. 3, 1989, pp. 583–609.

Pugh, Mrs. John. *Phrenology Considered in a Religious Light; Or, Thoughts and Readings Consequent on the Perusal of "Combe's Constitution of Man."* T. Ward and Co., 1846.

Pykett, Lyn. Introduction. *The Doctor's Wife*, Mary Elizabeth Braddon, edited by Lyn Pykett, Oxford UP, 2008.

Rauch, Alan. *Useful Knowledge: The Victorians, Morality, and the March of Intellect*. Duke UP, 2001.

"Recent Attacks on Phrenology." *Phrenological Journal*, vol. 11, no. 56, 1838, pp. 260–66.

"Remarkable Case of Change of Character and Pursuits, with Corresponding Change in the Form of the Head." *Phrenological Journal*, vol. 13, no. 65, 1840, pp. 341–44.

"Report of the Proceedings of the Phrenological Association." *Phrenological Journal*, vol. 15, no. 73, 1842, pp. 291–318.

"Rev. Antoinette Brown Blackwell." *Phrenological Journal and Science of Health*, vol. 103, no. 4, 1897, pp. 158–60.

Review of *The Tenant of Wildfell Hall*. *Sharpe's London Magazine*, vol. 7, 1848, pp. 181–84.

Richards, Evelleen. "Darwin and the Descent of Woman." *The Wider Domain of Evolutionary Thought*, edited by David Oldroyd and Ian Langham, D. Ridel Publishing, 1983, pp. 57–111.

———. "Huxley and Woman's Place in Science: The 'Woman Question' and the Control of Victorian Anthropology." *History, Humanity and Evolution: Essays for John C. Greene*, edited by James R. Moore, Cambridge UP, 1989, pp. 253–84.

Rippon, Gina. *The Gendered Brain: The New Neuroscience that Shatters the Myth of the Female Brain*. Vintage, 2019.

Roberts, Caroline. *The Woman and the Hour: Harriet Martineau and Victorian Ideologies*. U of Toronto P, 2002.

Romanes, George J. "Concerning Women." *The Forum*, vol. 4, 1887, pp. 509–18.

Rose, Nikolas. *The Politics of Life Itself: Biomedicine, Power, and Subjectivity in the Twenty-First Century*. Princeton UP, 2006.

Russett, Cynthia Eagle. *Sexual Science: The Victorian Construction of Womanhood.* Harvard UP, 1989.
Ryall, Anka. "Medical Body and Lived Experience: The Case of Harriet Martineau." *Mosaic,* vol. 33, no. 4, 2000, pp. 35–53.
Rylance, Rick. *Victorian Psychology and British Culture.* Oxford UP, 2000.
[Saunders, Emily Susan Goulding]. *Christian Phrenology.* Gustav Cohen, 1881.
———. "Dr. Spurzheim." *Ladies Magazine and Literary Gazette,* vol. 5, 1832, pp. 570–72.
Sax, Leonard. *Why Gender Matters: What Parents and Teachers Need to Know About the Emerging Science of Sex Difference.* 2nd ed., Harmony, 2017.
Schaub, Melissa. "A 'Divine Right to Happiness': The Sublime, the Beautiful, and the Woman Reader in *Madame Bovary* and *The Doctor's Wife.*" *Nineteenth-Century Feminisms,* vol. 7, 2003, pp. 23–39.
Schiebinger, Londa. *The Mind Has No Sex? Women in the Origins of Modern Science.* Harvard UP, 1989.
Seys, Madeline. "The Scenery and Dresses of her Dreams: Reading and Reflecting (on) the Victorian Heroine in M. E. Braddon's *The Doctor's Wife.*" *Changing the Victorian Subject,* edited by Maggie Tonkin, Madeleine Seys, and Sharon Crozier-De Rosa, U of Adelaide P, 2014, pp. 177–200.
Shapin, Stephen. "Phrenological Knowledge and the Social Strecture of Early Nineteenth-century Edinburgh." *Annals of Science,* vol. 32, 1975, pp. 219–43.
Shaw, Margaret. "Narrative Surveillance and Social Control in *Villette.*" *SEL,* vol. 34, no. 4, 1994, pp. 813–33.
Shteir, Ann B. *Cultivating Women, Cultivating Science: Flora's Daughters and Botany in England, 1760–1860.* John Hopkins UP, 1996.
———. "Elegant Recreations? Configuring Science Writing for Women." *Victorian Science in Context,* edited by Bernard Lightman, U of Chicago P, 2008, pp. 236–55.
Shuttleworth, Sally. *Charlotte Brontë and Victorian Psychology.* Cambridge UP, 1996.
———. *George Eliot and Nineteenth-Century Science: The Make-Believe of a Beginning.* Cambridge UP, 1984.
———. "Introduction." *The Lifted Veil and Brother Jacob,* George Eliot, Penguin, 2001.
Simpson, James. *The Philosophy of Education, with its Practical Application to a System and Plan of Popular Education as a National Object.* 2nd ed., Adam and Charles Black, 1836.
Sinclair, May. "Introduction." *The Tenant of Wildfell Hall,* Anne Brontë, J. M. Dent and Sons, 1922.
Smart, Alexander. "On the Application of Phrenology in the Formation of Marriages." *Phrenological Journal,* vol. 8, no. 38, 1834, pp. 464–73.

Smiles, Samuel. *Self Help; with Illustrations of Character and Conduct.* John Murray, 1859.

Smith, Elsie Cassell. *The Amateur Phrenological Club. (Its Sayings and Doings). Phrenological Journal and Science of Health,* vol. 104, no. 5, 1897, pp. 230–33.

Smith, George. "Charlotte Brontë." *Cornhill Magazine,* vol. 9, 1900, pp. 778–95.

"Socialism, or the New Morality." *The Times,* 24 Oct. 1839, col. D, p. 6.

"South Sea Islanders." *Phrenological Journal and Miscellany,* vol. 2, 1825, pp. 239–40.

Sparks, Tabitha. "Fiction Becomes Her: Represenations of Female Character in Mary Braddon's *The Doctor's Wife.*" *Beyond Sensation: Mary Elizabeth Braddon in Context,* edited by Marlene Tromp, Pamela K. Gilbert, and Aeron Haynie, State U of New York P, 2000, pp. 197–209.

Spelke, Elizabeth S. "Sex Differences in Intrinsic Aptitude for Mathematics and Science?: A Critical Review." *American Psychologist,* vol. 60, no. 9, 2005, pp. 950–58.

Spencer, Herbert. "Psychology of the Sexes." *The Popular Science Monthly,* vol. 4, 1873, pp. 30–38.

Spillers, Hortense. "Moma's Baby, Papa's Maybe: An American Grammar Book." *Diacritics,* vol. 17, no. 2, 1987, pp. 64–81.

"Spinsters: Their Past, Present, and Future Work." *Phrenological Magazine,* 1880, pp. 109–15.

Spivak, Gayatri Chakravorty. *In Other Worlds: Essays in Cultural Politics.* Methuen, 1987.

———. "Three Women's Texts and a Critique of Imperialism." *Critical Inquiry,* vol. 12, no. 1, 1985, pp. 243–61.

Spurr, Mrs. Thomas. *Course of Lectures on the Physical, Intellectual, and Religious Education of Infant Children. Delivered before the Ladies of Sheffield.* George Ridge, 1836.

Spurzheim, G[asper]. *Outlines of Phrenology; Being also a Manual of Reference for the Marked Busts.* Treuttel, Wurtz, and Richter, 1829.

———. *Phrenology in Connexion with the Study of Physiognomy. Part I: Characters.* Treuttel, Wurtz, and Richter, 1826.

———. *A View of the Elementary Principles of Education: Founded on the Study of the Nature of Man.* Treuttel, Würtz, and Richter, 1828.

Stepan, Nancy. *The Idea of Race in Science: Great Britain, 1800–1960.* Archon Books, 1982.

Stern, Madeleine B. *Heads and Headlines: The Phrenological Fowlers.* U of Oklahoma P., 1971.

Strachan, J. 'The mapp'd out skulls of Scotia': *Blackwood's* and the Scottish Phrenological Controversy." *Print Culture and the Blackwood Tradition, 1805–1930,* edited by D Finkelstein, U of Toronto P, 2006, pp. 49–69.

Stiles, Anne. *Popular Fiction and Brain Science in the Late Nineteenth-Century.* Cambridge: Cambridge UP, 2012.

"Summary of the Proceedings of the Edinburgh Philosophical Association." *The Philosophy of Education: With Its Practical Application to a System and Plan of Popular Education as a National Object* by James Simpson. 2nd ed., Adam and Charles Black, 1836, pp. 238–49.

Summers, Lawrence H. "Remarks at NBER Conference on Diversifying the Science and Engineering Workforce." *Harvard University*, 14 Jan. 2005. www.harvard.edu/president/speech/2005/remarks-nber-conference-on-diversifying-science-engineering-workforce

Swenson, Kristine. "Phrenology as Neurodiversity: The Fowlers and Modern Brain Disorder." *Progress and Pathology: Medicine and Culture in the Nineteenth Century*, edited by Melissa Dickson, Emilie Taylor Brown, and Sally Shuttleworth, Manchester UP, 2020, pp. 99–123.

T.A.F. "Mrs. J. Ellen Foster." *Phrenological Journal and Science of Health*, vol. 106, no. 5, 1898, pp. 148–49.

Thormählen, Marianne. "The Villain of *Wildfell Hall*: Aspects and Prospects of Arthur Huntingdon," *Modern Language Review*, vol. 88, no. 4, 1999, pp. 831–41.

Tomaiuolo, Saverio. *In Lady Audley's Shadow: Mary Elizabeth Braddon and Victorian Literary Genres*. Edinburgh UP, 2010.

Tomlinson, Stephen. *Head Masters: Phrenology, Secular Education, and Nineteenth-Century Social Thought*. U of Alabama P, 2005.

Torgerson, Beth. *Reading the Brontë Body: Disease, Desire, and the Constraints of Culture*. Palgrave, 2005.

"To the Editor of the *London Medical Gazette*." *London Medical Gazette*, Vol. 5, 1830, pp. 826–27.

Uglow, Jennifer. *George Eliot*, Pantheon Books, 1987.

Vogt, Carl. *Lectures on Man: His Place in Creation, and in the History of the Earth*. Longman, Green, Longman, and Roberts, 1864.

Wagner, Tamara S. "Speculations on Inheritance and Anne Brontë's Legacy for the Victorian Custody Novel." *Women's Writing*, vol. 14, no. 1, 2007, pp. 117–39.

———. "'We have orphans [. . .] in stock': Crime and the Consumption of Sensational Children." *The Nineteenth-century Child and Consumer Culture*, edited by Dennis Denisoff, Ashgate, 2008.

Wallerstein, Mrs. Henry. "Child Study and Phrenology." *Phrenological Journal and Science of Health*, vol. 107, no. 2, 1899, pp. 55–56.

Wallis, Roy. Editor. *On the Margins of Science: The Social Construction of Rejected Knowledge*, University of Keele, 1979.

Warne, Joseph A. *Phrenology in the Family. Or, The Utility of Phrenology in Early Domestic Education*, Maclachlan, Sewart, and Co., 1843.

Warwick, Alex. "Margins and Centers." *Repositioning Victorian Sciences: Shifting Centres in Nineteenth-Century Scientific Thinking*, edited by David Clifford et al., Anthem P, 2006, pp. 1–13.

Waters, Sarah. Introduction. *The Trail of the Serpent*, Mary Elizabeth Braddon. Edited by Chris Willis, Random House, 2003, pp. xv–xxiv.
Watson, Hewett C. *Statistics of Phrenology: Being a Sketch of the Progress and Present State of that Science in the British Islands*. Longman, Rees, Orme, Brown, Green, and Longman, 1836.
Winter, Alison. "The Construction of Orthodoxies and Heterodoxies in the Early Victorian Life Sciences." *Victorian Science in Context*, edited by Bernard Lightman, U of Chicago P, 2008, pp. 24–50.
———. *Mesmerized: Powers of Mind in Victorian Britain*. U of Chicago P, 1998.
Webb, R. K. *Harriet Martineau: A Radical Victorian*, Columbia UP, 1960.
Wells, R. B. D. *A Delineation of the Character, Talents, Physiological Developments and Natural Adaptations of [. . .] as given by [. . .]*. J. Burns, 1878.
Wheeler, Anna. "Rights of Women." *British Co-operator*, vol. 1, no. 1, 1830, pp. 12–15.
———. "Rights of Women." *British Co-operator*, vol. 1, no. 2, 1830, pp. 33–36.
"Women's Brains." *Phrenological Magazine*, vol. 4, no. 42, 1888, pp. 232–33.
"Women Physicians in the United Kingdom." *Phrenological Magazine*, vol. 4. No. 45, Sept. 1888, pp. 387–88.
"Where the Brontës Borrowed Books." *Brontë Society Transactions*, vol. 11, no. 5, 1968, 344–58.
Willis, Chris. Afterword. *The Trail of the Serpent*, Mary Elizabeth Braddon, edited by Chris Willis, Random House, 2003, pp. 408–14.
Wilson, John. "Noctes Amrosianae." *Blackwood's Edinburgh Magazine*, vol. 22, 1827, pp. 105–34.
Wise, Thomas James, and John Alex Symington, eds. *The Brontës: Their Lives, Friendships and Correspondence*, vol. 3, Shakespeare, 1932.
Wolff, Robert Lee. "Devoted Disciple: The Letters of Mary Elizabeth Braddon to Sir Edward Bulwer-Lytton, 1862–1873." *Harvard Library Bulletin*, vol. 22, 1974, pp. 5–35, 129–61.
———. *Sensational Victorian: The Life and Fiction of Mary Elizabeth Braddon*. Garland Publishing, 1979.
Woods, Derek. "Sanitation and Telepathy: George Eliot's *The Lifted Veil*." *Victorian Literature and Culture*, vol. 45, no. 1, 2017, 55–76.
Wordsworth, William. *The Prelude. William Wordsworth: The Major Works*, Oxford UP, 2000.
Wright, T. R. "From Bumps to Morals: the Phrenological Background to George Eliot's Moral Framework." *Review of English Studies*, vol. 33, no. 129, 1982, 35–46.
Xiao, Yizhi. "Lost in Magnification: Nineteenth-Century Microscopy and 'The Lifted Veil.'" *George Eliot–George Henry Lewes Studies*, vol. 69, no. 1, 2017, pp. 68–88.

Index

alcoholism, 92–94, 216n4, 219–20nn30–33
Allan, J. McGriggor, 28, 29–30, 40
American Institute of Phrenology, 35
Anthony, Susan B., 37
Anthropological Society of London, 28
Armstrong, Nancy, 60–61, 78, 216n2
atheism, 18–19, 80, 99–100, 107, 114, 123–24
Atkinson, Henry George, 17, 75, 99–100, 220n1, 221n7, 221n10, 222n14, 222n18; *The Letters on the Laws of Man's Nature and Development* (with Martineau), 100, 105, 106–17, 121–28
attraction. *See* physical attraction
Augustine of Hippo, 102, 105, 111–13, 115, 117–18, 120–21, 127, 222n23
authority: of authors and narrative, 192–93; of God, 72; of human character, 151; of husbands, 74; of nature, 61, 72–73; scientific, 7, 35, 151, 212n14, 227n3; of women, 90, 114–15, 192

Bain, Alexander, 10, 13
Barker, Juliet, 65, 217nn12–13
Baron-Cohen, Simon, 205
Barwell, Louisa, 90

beauty and physical attractiveness: Brontë, Charlotte, on "lack of," 65, 217nn12–13; in *Daniel Deronda*, 173, 176, 184–85, 188–89, 191–95; evolutionary anthropologists on, 28; in *Jane Eyre*, 65–73; phrenology and, 76–79; in *The Tenant of Wildfell Hall*, 76. *See also* physical attraction
Beauvoir, Simone de, 5, 200
Becker, Ernst, 169
Becker, Lydia, 39–40, 42, 43, 215n12
bias, 8, 25; ideological bias, 190, 191, 193, 196–97; observational bias, 57; selection bias, 26
Blackwell, Antoinette Brown, 52–53
Blackwood's Magazine, 31, 63, 129, 165–66, 175, 216n6
Bodenheimer, Rosemarie, 192
booklets, phrenological delineation, 31, 45, 75, 215n15
Bordo, Susan, 5
Braddon, Mary Elizabeth, 4, 17, 19, 130; *Aurora Floyd*, 131; *Lady Audley's Secret*, 131, 143, 147; self-defense against *Blackwood's Magazine* criticism, 129–30. *See also* Braddon, Mary Elizabeth, *The Doctor's Wife*; Braddon, Mary Elizabeth, *The Trail of the Serpent*

249

Braddon, Mary Elizabeth, *The Doctor's Wife*, 148–63; character in, 150–155; education in, 154, 156–60; influence of, 148; male privilege in, 149–50; plot overview, 148; publication and reception of, 148; self-knowledge in, 153; social categories in, 161–62; women's opportunities in, 152–54, 158

Braddon, Mary Elizabeth, *The Trail of the Serpent*, 130, 131–47; character in, 130–35, 140–46; as first British detective novel, 131, 233n1; institutional power in, 132–33, 137–47; phrenological reading in, 19, 133–34; physiological knowledge in, 131, 145–47; plot overview, 131–32; publication and reception of, 131, 147; publication of, 131; self-knowledge in, 110, 112, 115, 119, 121–22

brain: addiction and, 92–93, 219n31; anterior lobe, 166, 177, 180, 185–87; children's brains, 90–91; contemporary neuroscience, 203–9; craniometry, 16, 26, 29, 56–58; dissection, 12, 35, 220n2; frontal lobe, 44, 68, 167, 180; imaging technology, 203–4; "inferior" and "superior" brains by ethnicity, 69–71, 85–86, 168, 218n17; Martineau's bequest of her brain to science, 99, 220nn1–2; occupations and, 44, 53; physiological structural similarity by sex, 36–40, 42–43, 49–50, 52; plasticity and malleability of, 24–25, 176, 208–9; size and weight differences by sex, 23–28, 40, 213n1. *See also* phrenological readings

Bray, Charles, 17, 150–52, 157, 165–66, 169, 178, 226n17

British Association for the Advancement of Science, 33, 39–40

British Phrenological Association, 35, 49, 107–8

Broca, Paul, 10, 16, 26–27, 58, 213n1

Brontë, Anne, 61–62; *Agnes Grey*, 73–74. *See also* Brontë, Anne, *The Tenant of Wildfell Hall*

Brontë, Anne, *The Tenant of Wildfell Hall*, 18, 61, 73–97; ideology of feminine influence in, 74, 76, 78–83, 91–92, 93, 95–96, 219n25; marriage in, 74–82; phrenology in, 216n4; physical attraction in, 76, 77–78, 79; progressive positions in, 74; reviews of, 218n19; spousal compatibility in, 73, 75–77

Brontë, Charlotte, 61–62; on *Letters on the Laws of Man's Nature and Development*, 123; phrenological reading of, 17, 63–65; physical appearance of, 64–66; *The Professor*, 83–91; *Villette*, 216n2, 219n28. *See also* Brontë, Charlotte, *Jane Eyre*

Brontë, Charlotte, *Jane Eyre*: beauty and physical attractiveness in, 65–73; ideology of feminine influence in, 71–73; spousal compatibility in, 65–67; wealth in, 65–66, 67, 71, 73

Brontë, Patrick, 62, 63, 123, 125, 223n29

Brooks, Peter, 111, 115

Browne, J. P., 63–64

Bulwer-Lytton, Edward, 129, 148, 150

Bunyan, John, 118

Burns, Robert, 38

Bushnan, J. Stevenson, 114

Butler, Judith, 5

Caldwell, Charles, 92, 94, 219–20nn31–33
Cantlon, Jessica, 208
Capuano, Peter, 187–88
Carpenter, Mary Wilson, 184
Chapman, John, 177–78, 221n8, 223n25
Chappelsmith, Mrs., 75
Cheyette, Bryan, 187, 188, 228n21
Christianity: Augustine of Hippo, 102, 105, 111–13, 115, 117–18, 120–21, 127, 222n23; in *Daniel Deronda*, 181, 189–91, 200; in *Jane Eyre*, 72; in *Letters on the Laws of Man's Nature and Development*, 110–13, 116–18, 120–21, 123, 127–28; Paul the Apostle, 105, 111, 117–18, 125, 127; phrenology and, 17, 31, 85, 96, 218n17; in *The Tenant of Wildfell Hall*, 79–80, 96
Cixous, Hélène, 5
class: in *The Doctor's Wife*, 152–58, 162–63; in *Jane Eyre*, 71, 73, 97; phrenology and, 136, 152–53; in *The Professor*, 87; in *The Tenant of Wildfell Hall*, 78, 92, 97; in *The Trail of the Serpent*, 136, 143, 146–47
Cobbe, Frances Power, 23–26, 36
colonialism, 15, 62, 86
Combe, George: *The Constitution of Man in Relation to External Objects*, 12, 13–14, 54, 70, 72, 85, 93, 169, 178, 213n17, 218n17, 219n25; "Criminal Legislation and Prison Discipline," 177–78; *Elements of Phrenology*, 38, 173; *Moral Philosophy*, 67, 75, 77; on "natural" hierarchies, 177–78; personal relationships with women writers, 17; *System of Phrenology*, 85; women allowed in lectures by, 33–34
conduct manuals, 50, 60
confession and confessional narrative: blush as somatic confession, 140–41; *Letters on the Laws of Man's Nature and Development* and, 18, 101–5, 110–18, 121, 123–24, 126, 128
constructivism, 5–6
conversion, religious narratives, 104–5, 110–11, 117–20, 127–28, 224n5
conversion, secular phrenological, 18–19, 103–5, 127–28
Cooter, Roger, 45–46, 48, 80, 125, 214n6, 217n14
Courvoisier, François, 136
craniometry, 16, 26, 29, 56–58
Crawford, Emily, 44
crime and criminality: Combe on phrenology to prevent crime, 177–78; Galton on "criminal type," 186–87; proto-eugenic phrenological aims and, 15; in *The Trail of the Serpent*, 19, 131–47, 225n13

Dallas, E. S., 10
Dames, Nicholas, 10, 64, 212n13, 215nn1–2
Daniel Deronda. See Eliot, George, *Daniel Deronda*
Darwin, Charles, 12; *The Descent of Man*, 8, 27, 41–42; on education of women, 16, 27–28, 213n2; on mental differences between men and women, 8, 16, 24, 27–28, 41–42, 58; *Origin of Species*, 12
deconstructive turn, in feminist studies, 21

determinism: in *Daniel Deronda*, 181–82, 189, 192–94; free will and, 17, 80, 94–95, 123; in *Jane Eyre*, 70, 73, 218–19n23; in "The Lifted Veil," 19–20, 173–76, 197; in "A Minor Prophet," 167, 181, 197; phrenology and, 1–2, 4, 6, 10–11, 12, 14–15, 16–17, 24–25, 38, 43, 48, 51, 56–58, 62, 151, 219n25; in *The Tenant of Wildfell Hall*, 18, 81–82, 94, 97, 219n30; in *The Trail of the Serpent*, 141. *See also* essentialism

Deville, James, 134, 155, 185, 224n5, 226n20, 227n2

Dickens, Charles, 148, 215n1, 223n1

Diprose, Rosalyn, 211n3

Doctor's Wife, The. *See* Braddon, Mary Elizabeth, *The Doctor's Wife*

Donovan, Cornelius, 166, 178–79, 227n2

Downing, Harriet, 63, 216–17n8

Doyle, Arthur Conan, 44

Ecker, Alexander, 28

Ecker, Lydia, 39–40, 42–43, 215n12

Edinburgh Philosophical Association, 33

education: contemporary neuroscience and, 205–6, 208; in *The Doctor's Wife*, 154, 156–60; evolutionary biology on, 16, 27–28, 213n2; in "The Lifted Veil," 167, 172–75; national system of, 88, 169–70; phrenology and, 9, 12, 14, 32–33, 36–37, 39, 43, 52–54, 62–63, 83–91, 151–52, 169–70, 172; in *The Professor*, 83–91; reform, 169–72; in *The Tenant of Wildfell Hall*, 74, 90–91; women's barriers to, 32–33, 36–37, 39

Eliot, George: critical of phrenology as predictive for populations, 177–78; on education, 168, 197; interest in phrenology, 165–67; The Lifted Veil," 19–20, 167, 172–83, 186, 196, 197, 227–28n11, 228n15; *Middlemarch*, 148; "A Minor Prophet," 20, 167, 180–82, 185, 196–97; phrenological readings of, 166–67; "Woman in France," 168. *See also* Eliot, George, *Daniel Deronda*

Eliot, George, *Daniel Deronda*, 20, 167–68, 181–201; Christianity in, 181, 189–91, 200; Judaism in, 20, 182, 185–96, 199–201; "The Lifted Veil" and, 181–83, 186, 196, 197; "A Minor Prophet" and, 181–82, 185, 196, 197; race in, 20, 167–68, 185, 189–93, 196–97, 201; receptivity in, 187, 194–95, 228n21; vision of the future in, 181–201

Eliot, Lise, 207

Elliotson, John, 80, 107

Ellis, Ida Mitchell, 31

Ellis, Sarah, 60

Engledue, William, 107–9, 222n14

English Woman's Journal, The, 33

Englishwoman's Review, The, 23–24, 38

essentialism: contemporary neuroscience and, 204–5, 208–9; in *Daniel Deronda*, 20, 167–69, 193–94, 197–201, 203; in *Jane Eyre*, 73; in *Letters on the Laws of Man's Nature*, 19, 120–21, 128; phrenology and, 4–5, 9, 19, 20, 30, 38, 43, 49, 51–52, 56, 135; in *The Tenant of Wildfell Hall*, 61, 74, 81, 97, 219n30; in *The Trail*

of the Serpent, 143–46. See also determinism
Ethnological Journal, The, 85
Ethnological Society, 28, 33
eugenics: Galton, Francis, and, 70, 185–87, 197, 228n17; phrenology and, 15; spiritual eugenics, 196
evolutionary anthropology, 16, 24–29, 43, 58, 205–6, 212n12
evolutionary psychology, 25–26, 58, 165, 206

Fairman, Mrs. James, 54–56
Fawcett, C. D., 57, 58
Fee, Elizabeth, 57
feminine domestic influence, ideology of: in *Jane Eyre*, 71–73; phrenology and, 60–61; in *The Tenant of Wildfell Hall*, 74, 76, 78–83, 91–92, 93, 95–96, 219n25
feminist phrenology, 16–17, 20, 23–58; craniometry and, 26, 29; evolutionary theorists and, 26–29; origins and early history of, 30–35; personal identity and, 45–50; physiological psychology and, 35–45; women's access and integration, 32–36; women's mental abilities and, 36–45. See also phrenology and phrenological principles
Ferrier, David, 10
Fine, Cordelia, 207, 229n2
Flaubert, Gustave, 148
Flint, Kate, 160, 162, 226n22
Foucault, Michel, 17, 111–12, 114–15, 130–31, 137, 140, 146, 149, 215–16n2
Fowler, Jessie Allen, 35, 42, 49, 52–53, 214n11, 215n14
Fowler, Lorenzo Niles, 2, 35, 214n11

Fowler, Lydia, 37–38, 214n11
Fowler Institute, 35
free will, 17, 80, 94–95, 123. See also determinism
Friedan, Betty, 161–62
Froude, James Anthony, 128
Fuller, Margaret, 37
Fuss, Diana, 20, 194

Gage, Matilda Joslyn, 37
Gall, Franz Joseph, 11–13, 103, 166, 180, 204, 213n20
Galton, Francis, 70, 185–87, 197, 228n17
Gaskell, Elizabeth, 217nn12–13
Gates, Barbara T., 11
Geological Society, 33
Gieryn, Thomas, 12
Gilman, Sander, 190, 228n19
Ginzburg, Carlo, 186–87
Gould, Stephen Jay, 25, 26, 213n1
Graham, Thomas, 63
Grosz, Elizabeth, 5
Gurian, Michael, 204–5

Hall, Mary Lyon, 52
Hall, Spencer T., 105–6
Hamilton, Mrs., 1–4
happiness, 54, 127, 155–56, 172, 174
Haraway, Donna, 9
Hardy, Thomas, 148
Hawkins, John Isaac, 108
Hazard, Caroline, 53–54
Herschel, Caroline Lucretia, 52
Herschel, John, 2, 211n1
Herschel, William, 52
Hoecker-Drysdale, Susan, 105, 114, 222n18
Holyoake, G. J., 127
Hunt, Harriot, 33, 37
Hunter, Shelagh, 220n4

Huxley, Thomas Henry, 16, 24, 27–28, 33
imaging technology, 203–4
imperialism, 15, 62, 86
insanity, 70, 93, 132, 224n5
Irigaray, Luce, 5
Isis (journal), 31

Jackson, John Hughlings, 10
James, Henry, 187
Jane Eyre. See Brontë, Charlotte, *Jane Eyre*
Jeffrey, Francis, 125
Joel, Daphana, 208
Jones, Ann Rosalind, 5
Jordan-Young, Rebecca M., 207
Joshi, Priti, 74
Judaism and Jewish identity: in *Daniel Deronda*, 20, 182, 185–96, 199–201; in "The Lifted Veil," 182

Keighley Mechanics' Institute, 62–63, 217n10
Kersey, Alyssa, 208
Knoepflmacher, U. C., 187–88
Know Thyself (journal), 31
"Know Thyself" (phrenological motto), 48–49
Kristeva, Julia, 5
Kurshan, Ilana, 11, 105, 212n14

Ladies Magazine, 34
Lancet, The, 40, 136
Laqueur, Thomas, 6, 36, 211n4
Lee, Alice, 57–58
Letters on the Laws of Man's Nature and Development. See Martineau, Harriet, *Letters on the Laws of Man's Nature and Development*
Levine, George, 187, 188, 228n21
Lewenz, Marie, 57–58

Lewes, George Henry, 2, 10, 13, 151, 165–66, 183, 211n1, 213n20, 227nn3–4
Literary Gazette, 34
London Anthropological Society, 28
London Medical Gazette, 30–31
Lyons, J. C., 67

Macnish, Robert, 63, 92, 104, 217n10, 220n33
madness, 70, 93, 132, 224n5
Manual of Phrenology, 62
marriage: compatibility, 14, 18, 59–61, 65–67, 73, 75–77; gendered division of public and private spheres, 60–61; phrenology and, 67–68, 74–82. See also physical attraction
Martineau, Harriet: anti-theism of, 100, 102, 107, 222n16; atheism and agnosticism of, 99, 100, 107, 114, 123; *Autobiography*, 100, 102, 103, 111–12, 124–25, 127; bequeathing skull and brain in will, 99, 220nn1–2; *Eastern Life, Present and Past*, 102, 103; "Female Industry," 220–21n5; *Letters on Mesmerism*, 106–7; persecution rhetoric of, 125–28; on publication and reception of *Letters*, 124–28; visit to phreno-mesmerist, 105–6. See also Martineau, Harriet, *Letters on the Laws of Man's Nature and Development*
Martineau, Harriet, *Letters on the Laws of Man's Nature and Development*, 110–28; Brontë, Charlotte, on, 123; Christianity in, 110–13, 116–18, 120–21, 123, 127–28; as confessional narrative, 18, 101–5, 110–18, 121, 123–24, 126, 128; deferential posturing

in, 113–14; epistolary form, 111; on hypocrisy, 112; on individual identity in, 122–26; mesmerism in, 105–7, 114, 120, 122–23; preface and purpose of, 113; public reaction to, 123–24; publication of, 124–25; rhetoric of conversion in, 117–20; rhetoric of submission in, 116–17
Martineau, James, 123–24
materialist phrenology, 4–6, 16–19, 80, 104–5
materiality and materialism: in *Daniel Deronda*, 184, 192–93, 195–97; in *The Doctor's Wife*, 130; evolutionary psychology and, 165; in *The Letters on the Laws of Man's Nature and Development*, 18–19, 99, 101, 104–13, 115–24, 127–28; phrenology and, 24–26, 42–44, 49–51, 53, 62–63, 80, 96–97, 107–10, 168–69, 216n7; in *The Tenant of Wildfell Hall*, 80, 96–97; in *The Trail of the Serpent*, 130
maternity and motherhood: evolutionary anthropologists on, 36; evolutionary biologists on, 26–27; feminist phrenologists on, 20, 30, 36–37, 49; natural abilities and, 198–99; role of mothers in child development, 90–91, 152, 219n25; separate spheres ideology and, 27; in *The Tenant of Wildfell Hall*, 91–96; women's bodies and biology, 36–37, 50; women's identity and, 20, 36–37, 49, 198–201
McKay, Brenda, 182
mesmerism, 100, 105–7, 122–23, 172, 178, 212n11, 221–22n12, 221n8; phreno-mesmerism, 105–6, 114, 120, 221n10

Miles, Mrs. L., 31–32
Mill, John Stewart: *The Subjection of Women*, 40–43; on women's suffrage, 28
mind/body dualism, 5–6
Moscucci, Ornella, 6
Moxon, Edward, 124, 223n25
Murphy, Patricia, 211–12n6

neuroscience, contemporary, 203–9
neurosexism, 207–8, 229n2
Novak, Daniel, 185, 187–88

O'Dell, Geelossapuss, 50
O'Dell, Stackpool E., 50
Oliphant, Margaret, 44, 129–30
otherness, 15

Patenall, Annie, 35, 48, 214n9
patriarchy, 4, 8–9, 20, 56, 198–99
Paul the Apostle, 105, 111, 117–18, 125, 127
Pearson, Karl, 57–58
Philosophical Transactions of the Royal Society, 57
Phrenological Journal and Science of Health, The, 35, 43–44, 52, 85, 104, 108, 114, 125, 135, 214n10
Phrenological Magazine, The, 24, 35, 48, 49–50, 52, 214nn9–10, 215n14
phrenological readings: behavioral mysteries and, 223–24n4; of Charlotte Brontë, 17, 63–64; in *The Doctor's Wife*, 161; gender neutrality of, 48; in *Jane Eyre*, 68–69; in "The Lifted Veil," 19–20, 167, 172–77; of public figures, 17, 44, 63–65; in *The Trail of the Serpent*, 19, 133–34
Phrenological Society of Edinburgh, 85

phrenology and phrenological principles: addiction and intemperance, 92–94; craniometry compared with, 26; criminal case studies, 134–36; definition of, 2; external influence, 1–2, 4, 14, 17, 39, 42, 76; lectures, 1–3, 12, 30–31, 33–35, 38, 49, 51, 63, 75, 90–91, 214n4, 227n3; marriage, 67–68, 74–82; mother's role in child's character development, 90–91, 219n25; phreno-mesmerism, 105–6, 114, 120, 221n10; physical attractiveness, 76–79; popularity of, 12–13; "predominate" and "preponderate," use of the terms, 218n21; profile of Pope Alexander VI, 84, 91; in rural communities, 62–63; schooling, 86–90. *See also* feminist phrenology; physiological psychology

phreno-mesmerism, 105–6, 114, 120, 221n10. *See also* mesmerism

physical attraction: evolutionary anthropologists on, 28–29; in *Jane Eyre*, 66–68, 70–71. *See also* beauty and physical attractiveness

physiological psychology: in *The Doctor's Wife*, 161–62; in *Jane Eyre*, 67–73; Lewes, George Henry, on, 211n1; Martineau, Harriet, and, 18–19, 100–101, 105, 109–10, 113, 119–20, 128; phrenology and, 1–4, 8–12, 35–37, 40–45, 48, 51–54, 203–4; in *The Trail of the Serpent*, 131–49, 225nn12–13; in *The Tenant of Wildfell Hall*, 75–77, 83, 92–97

Pike, Luke Owen, 28, 36
Pinker, Steven, 206, 228–29n1
Pitman, Isaac, 44
popular science writing, 7, 212n10

power: Foucault on, 130–31; knowledge systems and, 130–31; physiological knowledge and, 145–47; in *The Trail of the Serpent*, 132–33, 137–47

Prideaux, T. S., 108–9
professionalization of science, 7, 32–33, 212n8, 212n10
progressivism and progressive ideals: in *Daniel Deronda*, 168, 197–98; in *The Doctor's Wife*, 160, 161, 162; Eliot, George, on, 19; essentialism and, 6, 168, 197–98; evolutionary theory and, 27; knowledge and scientific inquiry, 116–17; Simpson, James, on, 170; in *The Tenant of Wildfell Hall*, 74; women's use of phrenology and, 9, 16, 17–18, 19, 30, 54

pseudoscience, 7
Pugh, Mr. John, 80
Pykett, Lyn, 148

race: in *Daniel Deronda*, 20, 167–68, 185, 189–93, 196–97, 201; in *Jane Eyre*, 70, 217–18n16; phrenology and, 4, 5, 15, 25, 51, 57, 69–70, 85–86, 178–80; in *The Professor*, 87

racism, 4, 20–21, 28, 58, 69, 86, 192
Rauch, Alan, 86, 216n2
religion, 18–19; case of innate disposition for religious career, 155–56; conversion, 104–5, 110–11, 117–20, 127–28, 224n5; knowledge and, 116–20; materialist phrenologists and, 79–80; pleasures of the body and, 120–21; sexual inequality and, 101–2; social types and, 224n5. *See also* atheism; Augustine of Hippo; Christianity; Judaism and Jews; theism

Index | 257

Richards, Evelleen, 213n2, 214n3
Rippon, Gina, 208–9
Romanes, George, 23–25, 27, 58
Rose, Nikolas, 204
Royal Horticultural Society, 33
Royal Society, 33
Russell, John, 169–70
Russett, Cynthia Eagle, 8, 54, 212n12
Rylance, Rick, 10

Sax, Leonard, 205
Schiebinger, Londa, 6–7, 32, 211n5
self-culture, 48, 53–54
self-discipline and self-control, 61–62, 71, 74
self-improvement, 48, 62–63, 80, 218–19n23
self-knowledge, 48–51, 101, 110, 112, 115, 119, 121–22, 153
self-sacrifice, 78–79, 81, 219n30
separate spheres ideology, 6–7, 16, 18, 27, 30, 36, 60, 96, 205
sexual attraction. *See* physical attraction
sexual complementarity, theory of, 16, 30, 61, 74
Shapin, Steven, 7
Sharples, Eliza, 31, 214n4
Shteri, Ann B., 212n10
Shuttleworth, Sally, 13, 17, 61, 62, 86–87, 194, 215–16n2, 216nn3–4, 218–19n23, 228n15
Simpson, James, 9, 93, 107–9, 173, 219n29; *Philosophy of Education*, 9, 88, 169–70
Sinclair, May, 74
slavery, 85, 86, 221n5
Smart, Alex, 59–60
Smiles, Samuel, 14, 142
Smith, Elsie Cassell, 51
Smith, George, 63–65

Somerville, Mary, 25, 52
Sorosis, 52
Spelke, Elizabeth, 208
Spencer, Herbert, 8, 10, 151, 165
Spillers, Hortense, 5
Spivak, Gayatri, 193–94, 199, 200, 217–18n16
Spurr, Mrs. Thomas, 90
Spurzheim, Johann Gaspar, 12, 34, 38, 70, 72, 85, 92–93, 103, 172, 173–74, 214n7, 219n26
Stanton, Elizabeth Cady, 37
STEM, 206, 208
Stepan, Nancy, 15, 86
stereotypes, 54, 190–91, 201
Stiles, Anne, 10
subjugation of women, 1–2, 29–30, 60–61, 112, 149, 200, 211n3
Summers, Lawrence, 206
Swenson, Kristine, 37

Tenant of Wildfell Hall, The. *See* Brontë, Anne, *The Tenant of Wildfell Hall*
theism, 19, 100–103, 105, 107, 108, 121; anti-theism, 100, 102, 107, 222n16
Thormählen, Marianne, 80, 92, 216n4
Tomlinson, Stephen, 87
Trail of the Serpent, The. Braddon, Mary Elizabeth, *The Trail of the Serpent*

Universal Phrenological Society, 31
University College of London, 57–58

Vogt, Carl, 16, 24, 27, 28, 58, 213n2
Von Lersner, Elise, 90–91

Wagner, Tamara S., 216n4
Wallis, Roy, 7

Warwick, Alex, 7
wealth, 65–66, 67, 71, 73
Webb, R. K., 114
Wheeler, Anna, 38–39
Winter, Alison, 7, 35, 212n11
Wittgenstein, Ludwig, 186–87

Wolff, Robert Lee, 154, 163, 226n17
Wollstonecraft, Mary, 2
women's rights activists, 1–2, 16, 23, 29, 38, 58, 201
Wordsworth, William, 183–84
Wyse, Thomas, 169–70

www.ingramcontent.com/pod-product-compliance
Lightning Source LLC
Chambersburg PA
CBHW030533230426
43665CB00010B/869